Intense Medium Energy Sources of Strangeness
(UC–Santa Cruz, 1983)

TSIMESS

Theoretical
Symposium
on Intense
Medium
Energy
Sources
of Strangeness

at
UC–SANTA CRUZ
MARCH 19–21, 1983

AIP Conference Proceedings
Series Editor: Hugh C. Wolfe
Number 102
Subseries on Particles and Fields No. 31

Intense Medium Energy Sources of Strangeness

(UC–Santa Cruz, 1983)

Edited by
T. Goldman, H. E. Haber, H. F. -W. Sadrozinski
Santa Cruz Institute for Particle Physics

American Institute of Physics
New York 1983

Copying fees: The code at the bottom of the first page of each article in this volume gives the fee for each copy of the article made beyond the free copying permitted under the 1978 US Copyright Law. (See also the statement following "Copyright" below). This fee can be paid to the American Institute of Physics through the Copyright Clearance Center, Inc., Box 765, Schenectady, N.Y. 12301.

L.C. Catalog Card No. 83-72261
ISBN 0-88318-201-7

DOE CONF- 830339

FOREWORD

On March 18-21, 1983, some 90 Physicists gathered at the Santa Cruz campus of the University of California to discuss what particle physics questions should be addressed if much higher intensity, intermediate energy accelerators were to become available. The meeting was organized through the Santa Cruz Institute for Particle Physics (SCIPP) and supported by SCIPP, and by the National Science Foundation Office of Theoretical Physics, the U.S. Department of Energy Office of Intermediate Energy Physics, the UCSC Division of Natural Sciences, the Los Alamos National Laboratory Divisions of Physics and of Meson Physics, and the Tri-University Meson Facility of the National Science and Engineering Research Council of Canada. Most of the participants came from North America, but there were also several from Europe and a few from Japan.

It was the interest of the nuclear physics community in hypernuclei and in kaons as relatively weakly interacting hadronic probes of ordinary nuclei that initially prompted consideration of intermediate energy machines as "kaon factory" upgrades of their current pion factories. However, these proposed machines could well be extremely valuable to particle physicists. For this reason, we felt that careful thought on the relevant theoretical particle physics questions was warranted.

In the past few years, several workshops have been held at TRIUMF[1] and at Los Alamos[2] to discuss high intensity intermediate energy accelerators and the nuclear and particle physics experiments that might be done with such machines. The driving force was the realization that higher intensities offered new opportunities to search for rare processes and for very precise measurements of less rare ones. Some of the particle physics interests in rare decays and neutrino scattering were also considered at the Snowmass Workshop.[3] Although many potentially useful rare processes were discussed at those meetings, the members of the organizing and advisory committees of this conference were concerned that, at least for particle physics questions, the discussions had been driven too much by what could be done with particular machines. It seemed worthwhile to try a more theoretical approach of asking which studies of rare processes and precision measurements should be undertaken (if feasible) to advance our knowledge of fundamental physics. The conclusions would depend both on which predictions of the standard model were not yet tested as precisely as possible (or at all!) relative to others (i.e. -- consistent parameter determinations) and upon which were most likely to be observably affected by emendations reflecting additional (new) physics. The discussions that would go into formulating the conclusions would also provide a coherent basis for judging the relative value of proposed experiments. We believe that this volume goes a long way toward achieving these goals.

The Organizing Committee would like to thank all those who contributed to and participated in TSIMESS. A special note of thanks is due to the UCSC graduate students who provided transportation and other services, to Candi Arnott, Administrative Assistant to the Theory Group, and most especially to our Conference Secretary, Georgia Hamel of SCIPP, who really made everything happen.

Terry Goldman
Howard Haber
Hartmut Sadrozinski

REFERENCES

1. Proceedings of the Kaon Factory Workshop, TRIUMF report no. TRI-79-1, edited by M.K. Craddock (1979); Proceedings of the Second Kaon Factory Physics Workshop, TRIUMF report no. TRI-81-4, edited by R.M. Woloshyn and A. Strathdee (1981).
2. Proceedings of the Workshop on Nuclear and Particle Physics at Energies up to 31 GeV: New and Future Aspects, Los Alamos report no. LA-8775-C, edited by J.D. Bowman, L.S. Kisslinger and R.R. Silbar (1981). LAMPF II Workshop, Los Alamos report no. LA-9416-C, compiled by H.A. Thiessen (1982); Proceedings of the Second LAMPF II Workshop, Vol. I and II, Los Alamos report no. LA-9572-C, edited by H.A. Thiessen, T.S. Bhatia, R.D. Carlini and N. Hintz (1982).
3. Proceedings of the 1982 DPF Summer Study on Elementary Particle Physics and Future Facilities, Snowmass, Colorado, 28 June - 16 July 1982, edited by R. Donaldson, R. Gustafson and F. Paige.

ORGANIZING COMMITTEE

R. Brower

D.E. Dorfan

T. Goldman

H.E. Haber

J. Primack

H.F.-W. Sadrozinski

ADVISORY COMMITTEE

D. Bryman

M.K. Gaillard

F.J. Gilman

S.L. Glashow

C.M. Hoffman

R.E. Shrock

F. Wilczek

SPONSORED AND SUPPORTED BY:

University of California, Santa Cruz
Santa Cruz Institute for Particle Physics
Division of Natural Sciences

U. S. National Science Foundation
Office of Theoretical Physics

U. S. Department of Energy
Office of Intermediate Energy Physics

Los Alamos National Laboratory
Physics Division
Meson Physics Division

Tri-University Meson Facility of the National
Science and Engineering Research Council
of Canada

TABLE OF CONTENTS

INTRODUCTION

T. Goldman†

Welcome to the Theoretical Symposium on Intense Medium Energy Sources of Strangeness. The organizing and advisory committees would like to thank and acknowledge support for this conference from the National Science Foundation Office of Theoretical Physics, the Department of Energy Office of Intermediate Energy Physics, the Physics and Meson Physics Divisions of Los Alamos National Laboratory, and the Directors Office of the Tri-University Meson Facility in Vancouver. We are especially grateful to the Santa Cruz Institute for Particle Physics and to Dean Bill Doyle, Division of Natural Sciences for providing a substantial contribution on behalf of the University of California at Santa Cruz. Without the help of many people in these organizations, this symposium would not have been possible.

This meeting has come about due to the success of the low energy meson factories ($\lesssim 1$ GeV proton accelerators) at providing high quality, high intensity beams of secondary particles (muons and pions) for high data rate and high precision experiments. Many important physics problems accessible in this energy range have been (or soon will be) resolved. An outstanding example of this progress is the current upper limit of 1.9×10^{-10} for the branching ratio of the decay $\mu \rightarrow e\gamma$. The group working at Los Alamos expects to extend this search by more than two orders of magnitude by the end of this decade. I believe this is indicative of a certain maturity of the machines and of the experimental techniques. In view of this developing maturity, many experimentalists working at meson factories have begun to see that these facilities may soon have less to do. Consequently, they have started to look at what new problems will benefit from the application of the expertise they have acquired at handling intense particle beams.

Not surprisingly (to high energy physicists), they have considered the possibilities that arise when higher intensities are made available at somewhat higher energies. In the "medium energy" range ($\lesssim 30$ GeV), strange particles become readily available. This opens a vast new area of nuclear and hypernuclear physics to explore. Ben Gibson will give you an inkling in his talk of the many fascinating questions that can be raised.

Of course, strange particles are still of intrinsic interest themselves to particle physicists. Glashow's comment describing the study of very rare decays as one of the high energy frontiers is by now a commonplace. Just as studies of very rare processes involving muons and pions have been made possible by current meson factories, one can imagine similar developments at "kaon factories". It is also easy for us to believe this because of our experience with SPEAR and CESR as D and B meson "factories".

†Permanent address: Theoretical Division, Los Alamos National Lab, Los Alamos, NM 87545

However, I have been very bothered over the last few years, as I became involved in the issue of the value of such machines, by the way the questions were being put. The machine builders were confident of being able to produce the requisite machine. My experimental colleagues could then imagine doing this new experiment, or vastly improving that old one. And the question they would ask was: "Given that we can do this, what good would it be?"

While they were willing to take "None!" for an answer sometimes, it seemed to me, as a theorist, that they were asking the wrong question. Furthermore, many theorists have been and are continuing to ask the right (from our point of view) questions: Given what we know, suspect and conjecture, what things can we think of that will test our ideas?

Since this is not a totally new area of investigation, many of the answers will involve more precise measurements of familiar processes. But there will be new reasons, and now and again, new processes for study (at least in the sense that they may have been previously discarded as uninteresting or not worth the effort). Even if we end up with the same list of processes to study as our experimental colleagues have compiled, the exercise will have been worth the effort. We owe it to ourselves, and to our funding agencies, to determine as accurately and completely as possible, the value to particle physics of the proposed efforts to produce higher intensities at intermediate energies.

Most importantly, we must perform this evaluation independently of any particular proposed capability. We cannot afford to let the decision about some very expensive hardware be dominated by its feasibility and cost. We must decide first what is needed to advance the field.

LAMPF II

Cyrus M. Hoffman
Los Alamos National Laboratory, Los Alamos, NM 87545

ABSTRACT

A brief description of LAMPF II is presented. LAMPF II is a planned high intensity, 16-32 GeV proton accelerator with LAMPF as the injector. Some of the physics which could be studied with LAMPF II is discussed.

INTRODUCTION

Before I discuss the plans for LAMPF II, it seems appropriate to discuss some of the properties of "LAMPF I." LAMPF is 1/2 mile long linac which accelerates protons to a kinetic energy of 800 MeV. The most important property of LAMPF is the extremely high intensity of the beam. A typical intensity during the last several years was 650 μA (= 4 × 10^{15} protons/s). This is by far the highest intensity accelerator in this energy regime or at any higher energy. Three months ago, an intensity record of 1.2 mA was set: this is 20% above LAMPF's design specifications! At 650 μA, the average beam power is 0.5 MW. For comparison, if the Tevatron operates at 10^{12} protons/s (average), the average beam power there would be 0.16 MW.

The intense proton beam at LAMPF is used to produce intense, high quality beams of $\pi^{\pm}$, $\mu^{\pm}$, n and ν_e for particle and nuclear physics research. In addition, the lower intensity primary beam, either polarized or unpolarized, makes possible a broad program of proton-scattering experiments. Finally, the primary beam is employed to produce a wide variety of radioisotopes and to perform radiation damage studies.

There are several improvements presently underway at LAMPF. At present, the intense proton beam is positively charged. A new H^- injector should be installed soon which will be the source for the 1 mA beam. An H^- beam simplifies injection into a circular machine. The second major improvement underway is the construction of the Proton Storage Ring (PSR). The PSR will produce very small duty factor beams for neutron-scattering studies. The PSR could also be used as a high-quality source of ν_μ and $\mu^{\pm}$ beams.

LAMPF II

a) The Accelerator

In thinking of a higher energy, high-intensity accelerator, it soon becomes clear that LAMPF is an ideal injector. First, the 800 MeV beam has a small emittance (σ = 0.05π mr-cm). This, plus the fact that the beam will be H^- simplify the problem of injection into a circular accelerator. The final energy of LAMPF is very high for an injector, thus reducing the effects of space-charge at injection. The high energy of LAMPF also implies that the rf frequency swing of a subsequent circular accelerator is no more than 19%. The rf system of this accelerator is probably the most critical component. The small rf frequency swing greatly simplifies the design of the rf system and will mean significant power savings.

LAMPF II is envisioned as a higher energy high-intensity machine. The decision as to what final energy should be chosen, ought to be dictated by the physics to be done there. Although the original motivation for considering LAMPF II was to produce kaons (a "kaon factory") it quickly became clear that LAMPF II would also be an extremely intense source of pions, muons, antiprotons, hyperons and neutrinos. Thus the term "Kaon Factory" is too narrow.

The design energy of LAMPF II has been motivated by the yields of secondary particles as a function of proton energy. Recently, a collaboration from Los Alamos, TRIUMF, CERN, Rome, Saclay, and Vanderbilt[1] systematically measured $\bar{p}$, K^-, and π^- production cross sections at 10, 18, and 24 GeV and 0°. As illustrated in the cross sections tend to rise faster than the proton energy. On the other hand, calculations[2] seem to indicate that neutrino fluxes are approximately proportional to the proton energy above ~2 GeV (see Fig. 2). However, the average neutrino energy does rise with proton energy so the neutrino event rate will rise faster than the proton energy. On the basis of these considerations, an energy of from 16 GeV to 32 GeV has been chosen as a reference design figure. I could imagine building LAMPF II as a 16 GeV machine and increasing the energy in later upgrades.

The intensity of LAMPF II will certainly be much lower than that at LAMPF. A design value of 100 μA (6×10^{14} protons/s) has been adopted. A 16 GeV beam with this intensity will have an average beam power of 1.6 MW. LAMPF II will be ~100 times more intense than past machines in this energy range.

In order to accelerate this many protons/s, LAMPF II will be a rapid cycling machine. Present design calls for 60 Hz: this implies a beam pulse of 10^{13} protons, equal to the number of protons in an AGS pulse. The beam will be fast extracted from the accelerator and injected into a DC stretcher ring occupying the same tunnel as the accelerator. Slow extraction from this ring will result in high-duty factor ($\gtrsim$ 80%) beams. Fast extracted beams will be used for neutrino and pulsed muon experiments. Table I shows a reference design for LAMPF II. We are contemplating using permanent magnets for the stretcher ring. Figure 3 shows the location of LAMPF II in relation to existing facilities.

b) Experimental Area

The fast extracted beams will be ν_μ, $\bar{\nu}_\mu$ and several muon beams. There is also a possible muon racetrack to produce ν_e and $\bar{\nu}_e$ beams.

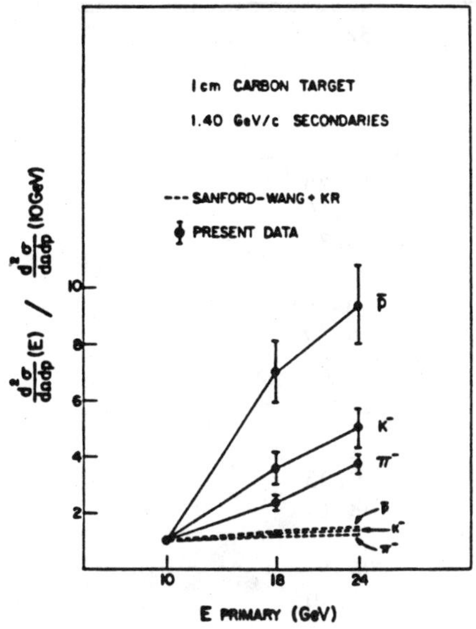

Fig. 1. Energy dependence of 0^0 production cross sections for $\bar{p}$, K^- and π^- on a 1-cm carbon target normalized to the cross section at 10 GeV (from Ref. 1). KR refers to kinematic reflection.

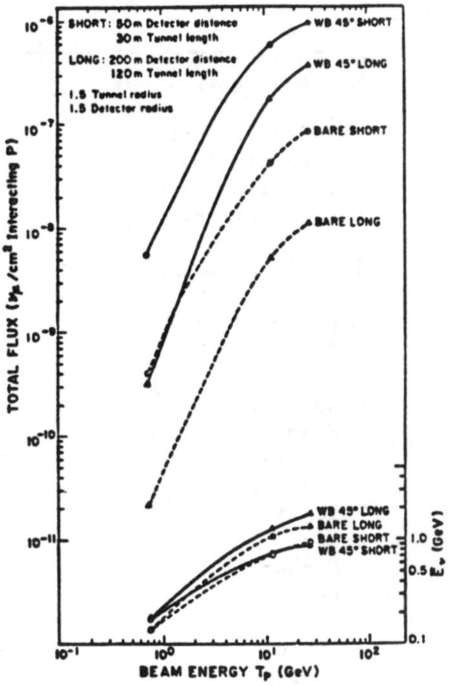

Fig. 2. Calculated ν_μ fluxes as a function of proton energy (from Ref. 2). The lower curves and the scale to the right show the average neutrino energy.

The high duty factor beams will include several separated beam lines. A 1.5 GeV/c K, π, and $\bar{p}$ beam will feed the HRS spectrometer while a lower momentum beam line (~400 MeV/c) will be used with the EPICS spectrometer. General purpose separated beams will be at ~400 MeV/c, ~800 MeV/c, ~1.5 GeV/c and up to 6 GeV/c. One or more neutral kaon beam lines will also be constructed. Finally, we envisage a time separated $\bar{p}$ beam,[3] ~1 Km long.

c) Costs

No accurate cost estimates have been made. Nevertheless, we guess that LAMPF II will cost ~\$75 M for the accelerator and stretcher and an additional \$75 M for experimental areas. The

TABLE I LAMPF II reference lattice

Maximum Energy, GeV	32
Injection Energy, GeV	0.8
Repetition Rate, Hz	60
Superperiodicity	4
Circumference, m	1011
Tune, Horizontal	17.44
Vertical	17.40
Transition Gamma	13.377
Maximum	
β_x, m	26.76
β_y, m	26.79
η_x, m	2.57
Quadrupoles	
Number	136
Length, m	1.4
Radius, m	0.05
Pole field, kG	8.13
Bends	
Number	96
Length, m	4.50
Angle, deg	3.75
Field, kG	15.973
Radius, m	68.75
Gap, m	0.05
Useful width, m	0.10
Entrance angle, deg	1.875
Exit angle, deg	1.875
Length of cells with bend, m	14.30
Cells without bend	20
Cell length, m	15.50
Drift per half cell, m	6.35

latter figure includes the beam lines, several large spectrometers, and two large, general purpose detectors.

d) Beam Line Design

A crucial question is how does one use all of this intensity? Generally, one trades in some intensity for beam quality as discussed below. On the other hand, there will be times when one may want as high a flux as possible to study very rare phenomena.

In present separated charged beams the ratio of kaons to pions is about 1/10. At LAMPF II we would construct beams with equal kaon and pion populations. The pions which survive in present beams come from particles which decay near the production target ("cloud" pions) and from pions scattering from pole pieces and

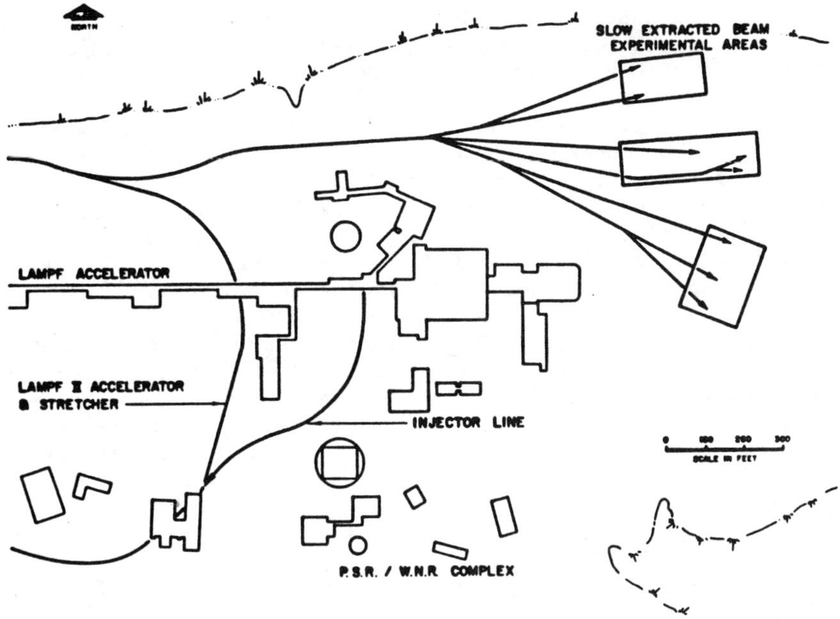

Fig. 3. Site plan for LAMPF II. The fast extracted areas are to the left of the figure.

slits. The need to keep the channel short in order to have a high enough kaon flux precludes the possibility of eliminating these pions. At LAMPF II better definition of the particle source, at the cost of increasing the channel length one kaon lifetime[4] will eliminate these pions.

K_L^0 beams at present machines have typically ten times more neutrons than kaons. At LAMPF II, one could preferentially absorb neutrons[5] utilizing the fact that $\sigma_{np}^{TOT} \simeq 40$ mb while $\sigma_{K_L^0 p}^{TOT} \simeq 24$ mb. A final example of improved quality beams at LAMPF II is the narrow band ν_μ beam. To study the y-dependence of neutrino-electron scattering, one needs a neutrino beam with well-defined energy.[6] Narrow-band beams at present facilities do not have a high enough flux to obtain a usable data sample. Such a beam at LAMPF II would be intense enough for such an experiment. In fact, the narrow-band beam at LAMPF II would be more intense than the wide-band beam at the AGS.

e) Physics at LAMPF II

The range of physics to be studied at LAMPF II includes exciting fundamental particle and nuclear physics. I will not discuss the particle physics as that would be prejudging the results of this workshop. I would like to briefly mention the nuclear phyics.

At LAMPF II, one would continue nuclear studies with pions (LAMPF II produces up to 100x more pions than LAMPF and enlarges the available energy range). In addition, one can begin precision studies of K^+-nucleus scattering. The K^+ is the cleanest hadronic probe of nuclear matter as the K^+-nucleon interaction is so (relatively) weak. At LAMPF II, a systematic study of hypernuclei could be undertaken using the very high intensity, clean kaon beams.

Finally, LAMPF II would permit greaty enhanced muon studies. Included here are rare decays (such as $\mu^+ \rightarrow e^+\gamma$), precision studies of muonium and the muon anamolous magnetic moment to further test QED, and μSR (muon spin rotation) studies in solid state physics.

For more details of all of these topics, I refer the reader to Ref. 7.

CONCLUSION

Interest in the physics to be addressed by LAMPF II is very high both within the nuclear physics community and amongst particle physicists. As we shall hear shortly, several other institutions are also contemplating a machine to do this physics.

The present schedule calls for submission of a formal proposal in 1984. The machine could be ready for experiments as early as 1992. An extremely rich and diverse physics program is in the offing.

REFERENCES

1. J. F. Amann et al., Los Alamos Report LA-9486-MS (1982).
2. K. C. Wang, R. C. Allen and H. H. Chen, University of California, Irvine Report, UCI-Neutrino #73 (1982).
3. T. E. Kalogeropoulos, "Antinucleon Time Separated Beams" in Proceedings of the Workshop on Nuclear and Particle Physics at Energies up to 31 GeV: New and Future Aspects, J. D. Bowman, L. S. Kisslinger and R. R. Silbar, Ed., LA-8775-C (1981).
4. P. Birien, CERN Report, CERN-EP/82-90 (1982); D. E. Lobb, University of Victoria Report VPN-81-5 (1981).
5. C. M. Hoffman, Los Alamos Report, LA-9337-MS (1982).
6. R. D. Carlini, "LAMPF II Neutrino Working Group Report" in Proceedings of the Second LAMPF II Workshop, H. A. Thiessen, T. S. Bhatia, R. D. Carlini and N. Hintz, Ed., LA-9572-C (1982).
7. Proceedings of the Workshop on Muon Science and Facilities at Los Alamos, R. H. Heffner, Ed., LA-9582-C (1982).

THE TRIUMF KAON FACTORY

Douglas Bryman
TRIUMF, 4004 Wesbrook Mall, Vancouver, B.C., Canada V6T 2A3

ABSTRACT

A kaon factory, using TRIUMF as an injector, could provide beams of kaons, neutrinos and other particles 100 to 1000 times more intense than are available at existing machines in the 30 GeV range. Consequently experimental opportunities would arise for new studies involving CP violation, rare kaon decays and fundamental neutrino interactions.

Due to the extraordinary success of the standard gauge model, encompassing the electroweak theory based on the SU(2)×U(1) and quantum chromodynamics based on SU(3), many low-energy experiments requiring measurements of extreme sensitivity and high precision have become important. Proton decay and double beta decay experiments and direct measurements of neutrino masses are representative examples. Many pion and muon experiments aimed at critically examining the predictions of the standard model have been made possible by the advent of the high-intensity, medium-energy facilities, LAMPF, SIN and TRIUMF.

The great intensities at the meson factories have been put to full use in experiments dealing with lepton number violation. Searches for direct transitions between lepton generations, involving lepton number nonconservation in reactions such as $\mu \to e\gamma$[1] and μ^-+Nucleus $\to e^-$+Nucleus[2], have been improved in sensitivity by two orders of magnitude in recent years. New experiments expect to reach between one and two orders of magnitude lower than current levels, approaching branching ratios of 10^{-12} relative to ordinary lepton number-conserving reactions.[3,4] Lepton flavor-changing reactions may be suppressed naturally by the leptonic analog of the Glashow-Iliopoulos-Maiani (GIM) mechanism, which suppresses the strangeness-changing neutral weak currents. They are sensitive to the existence of exotic superheavy neutral particles, such as leptons or additional Higgs scalars present in some theories, and might provide a window for observing effects related to ultra-high mass scales (~100 TeV).[5]

Another example is the study of muon decay, which has been the primary source of knowledge about the leptonic charged current. Measurements of the Michel parameters, which describe muon decay in terms of general interactions, are now in the process of being refined by factors of 10.[6] These experiments will place exacting constraints on unified theories. Deviations from the values of the Michel parameters predicted by the standard models would have far-reaching consequences. They could favor the alternative choice of left-right symmetric models, such as $SU(2)_L \times SU(2)_R \times U(1)_{L+R}$,[7] and indicate the existence of right-handed $W^{\pm}$ bosons with $M_{W_R} \gg M_{W_L}$.

Precise studies of pion decays also test the standard model in unique ways. Experiments have been undertaken to improve accuracies by a factor of ~5 in the decays $\pi \to e\nu$,[8] $\pi \to e\nu_e\gamma$,[9] and $\pi \to$

$\pi^0 e \nu_e$,[10] reaching levels at which the standard models can be severely tested. At TRIUMF a new measurement of the ratio $\pi \to e\nu_e/\pi \to \mu\nu_\mu$ has been completed which supports the principle of electron-muon universality at the 1% level. The decays $\pi^+ \to e^+\nu_e\gamma$ and $\pi^+ \to \pi^0 e^+\nu_e$ give sensitive information on the nature of pion structure and weak interaction couplings. A recent LAMPF measurement[11] of the decay rate for $\pi^0 \to e^+e^-$ has resulted in a factor of 3 improvement in accuracy and a new experiment[12] designed to measure the π^0 form factor with 10 times the precision of previous work is proposed at TRIUMF. Other exacting electroweak studies at the meson factories include: searches for neutrino oscillations and heavy neutrinos, measurement of the ν_μ mass, neutrino-electron elastic scattering and studies of the muonium system.

The successful development of the meson factories in the 0.5 to 0.8 GeV range with currents of 0.1 to 1 mA naturally leads to consideration of their use as injectors for high intensity, higher energy machines capable of in-depth attack on problems accessible only at higher energies. A proposal for a TRIUMF "kaon factory", an accelerator which would boost the intense beam to energies in the range 10^1 to 10^2 GeV, is now being developed. The advent of such a machine would provide new opportunities for the study of diverse topics, including CP violation, lepton flavor violation, weak neutral currents, the dynamics of quark systems beyond the first generation and many others.

Consider CP violation. In the superweak model,[13] observable CP violation is confined effectively to the neutral kaon system, since all other CP- or T-violating effects are at the 10^{-9} level or less. In milliweak models, such as the six quark Kobayashi-Maskawa model[14] or the Weinberg-Higgs model,[15] CP violating effects occur at the 10^{-3} level, compared to ordinary weak amplitudes, and manifestations of CP violations outside the neutral kaon system could exist. Examples of processes which could provide evidence to support milliweak models are $K^+ \to \pi^0\mu^+\nu_\mu$ and $K_L^0 \to \pi\mu\nu$, in which the presence of non-zero muon polarization transverse to the decay plane is an indicator of T-violation. Using their results on $K^+ \to \pi^0\mu^+\nu_\mu$ and $K_L^0 \to \pi\mu\nu$, Campbell et al.[16] found that the CP-violating muon polarization normal to the decay plane was $P_n = (-1.85 \pm 3.60) \times 10^{-3}$, a value which is somewhat below the prediction of the models where CP violation arises via Higgs exchange. To test alternative models of CP violation the accuracy of this and other such measurements should be improved by an order of magnitude. Intense, clean kaon beams developed at a kaon factory would make such improvements feasible.

In the neutral kaon system CP violation is conventionally parametrized in terms of the amplitudes for two-pion decays:

$$\frac{A(K_L^0 \to \pi^+ \pi^-)}{A(K_S^0 \to \pi^+ \pi^-)} = \varepsilon + \varepsilon'$$

and

$$\frac{A(K_L^0 \to \pi^0 \pi^0)}{A(K_S^0 \to \pi^0 \pi^0)} = \varepsilon - 2\varepsilon' \ .$$

Experiments, reaching the sensitivities of 10^{-3} to 10^{-4} in measurements of $|\epsilon'/\epsilon|$, may be required to choose between models.[17] Present experiments are at the percent level. High quality kaon beams with intensities far greater than those now available will be necessary to meet this challenge.

Other possible sources of potentially vital information on CP violation could come from high sensitivity studies of rare neutral current processes, such as $K \to \gamma\gamma$, $\mu^+\mu^-$ and πe^+e^-, which only occur in higher orders and at rates strongly suppressed by the GIM mechanism. These decays may provide particularly sensitive tests of milliweak models.[18] The expected branching ratios are extremely small ($<10^{-11}$) and measurements would require much more intense beams than are now available.

Observation of muon number violating K decays, such as $K_L^0 \to e^\pm\mu^\mp$, $K_S \to e^\pm\mu^\mp$, $K_L^0 \to \pi^0 e^+\mu^-$, and $K^+ \to \pi^+e^+\mu^-$, could add an extra dimension to the understanding of flavor-changing interactions. These processes, along with the ones mentioned earlier ($\mu \to e\gamma$, $\mu^- Z \to e^- Z$, etc.), may occur through neutral gauge boson exchange or scalar boson exchange, if flavor-changing Higgs particles exist, or through the existence of heavy neutral leptons. Assuming Higgs exchange, estimates for the branching ratio of $K_L^0 \to \mu^+e^-$ are in the range $\sim 10^{-12}$ compared to the present limit of 2×10^{-9}.[19]

Another very interesting rare process is the decay $K^+ \to \pi^+\nu\bar{\nu}$. $\nu\bar{\nu}$ pairs in the final state include any and all light neutrinos, so, in principle, the reaction rate could be an indicator of the number of neutrino flavors. Even though at present the rate may be calculable only to within an order of magnitude,[20] valuable information would result from a definitive measurement. The current experimental limit is 1.4×10^{-7}.[21] A K^+ beam, derived from a 30 GeV, 100 μA primary proton beam, would allow a measurement of a few per cent accuracy for a $K^+ \to \pi^+\nu\bar{\nu}$ branching ratio of 10^{-10}. In addition, such a measurement would allow an extremely sensitive search for axions (a), photinos ($\tilde{\gamma}$) and other exotic neutral particles through reactions such as $K^+ \to \pi^+ a$, $K^+ \to \pi\tilde{\gamma}\tilde{\gamma}$, etc.

There are many other areas involving the weak interactions of kaons which would merit further study using the intense beams potentially available at a K-factory. Some examples are determinations of Kobayashi-Maskawa mixing angles using K_{e3} decay, hyperon decays, and the $K_L^0 - K_S^0$ mass difference.[22] Studies of the structure dependence of the $\bar{s}d$ vertex in $K \to \mu\nu\gamma$ and $K \to e\nu\gamma$ decays, and experiments dealing with the $\Delta I = 1/2$ rule are other possibilities.

Searches for neutrino oscillation and measurements of neutrino-electron elastic scattering are further examples of experiments that would benefit immensely from intense high energy beams. Even if the Z^0 is discovered in the near future, neutrino electron scattering involving the reactions

$$\begin{gathered} \nu_\mu + e^- \to \nu_\mu + e^- \\ \bar{\nu}_\mu + e^- \to \bar{\nu}_\mu + e^- \\ \nu_e + e^- \to \nu_e + e^- \\ \bar{\nu}_e + e^- \to \bar{\nu}_e + e^- , \end{gathered}$$

can give unique information about the structure of the weak neutral current and about the interference between neutral and charged currents.[23] The cross sections are extremely small ($\sim 10^{-42}$ cm^2/GeV) and relatively few events ($<10^3$) of all types have been observed so far. The potential for intense neutrino beams at a K-factory could result in the first detailed studies of electron neutrino-electron scattering. With ν_e derived from neutral kaon beams via $K_L^0 \rightarrow \pi e \overset{(-)}{\nu}_e$, it is estimated that experiments with several thousand events could be achieved in a 100 day run, assuming a 30 GeV, 100 μA primary proton beam and a 10^3 ton detector. Moreover, detailed, high statistics, $\nu_\mu e \rightarrow \nu_\mu e$ studies could be done using narrow band beams.

At TRIUMF two approaches[24] are under active consideration for accelerating proton beams with sufficient energy for production of high flux beams of kaons, neutrinos and other particles with intensities two orders of magnitude or more greater than at existing facilities. The first approach is based on a pair of superconducting ring cyclotrons used to accelerate a 0.43 GeV TRIUMF beam to 3 GeV and then to 15 GeV. Proton currents up to ~400 μA are possible. The macroscopic duty factor would be 100%, as in the present operation at TRIUMF. Also the RF time structure of the beam would be similar to that of the present operation at 23 MHz and possibly could be put to great advantage in conjunction with RF particle separators to produce clean, charged kaon beams. Time-of-flight measurements for neutral beams would be another feature. A storage ring would be required to derive pulsed beams for neutrino experiments.

The second alternative involves the use of proton synchrotrons. In order to match the continuous time structure of the cyclotron to the discontinuous structure of the synchrotron, ~100 turns could be stacked in the cyclotron and then repeatedly injected into the synchrotron. Presently designs are being considered to reach 15 to 30 GeV. The designs involve a rapid cycling (15 to 30 Hz) synchrotron which would operate at 100 μA with duty factors of $\sim 10^{-5}$ and 50% under different modes of extraction. Low duty factor operation would be an important feature for neutrino experiments. Figure 1 shows a possible layout for a 30 GeV synchrotron. Some specifications for a 16 GeV synchrotron are listed in Table I. Other synchrotron designs leading to energies in the range 30-60 GeV are being considered as well.

In conclusion, serious consideration is now being given to a high intensity accelerator which would boost TRIUMF's intense beam to energies of 10^1 to 10^2 GeV. This facility would be much more than just a "kaon factory", since beams of pions, muons, protons, antiprotons, hyperons, and neutrinos would be available with intensities 100 to 1000 times greater than those at existing accelerators. As was the case for the meson factories, a new and exciting era in the development of high precision experiments can be anticipated.

ACKNOWLEDGEMENT

I would like to thank E.W. Blackmore for providing the layout shown in Fig. 1.

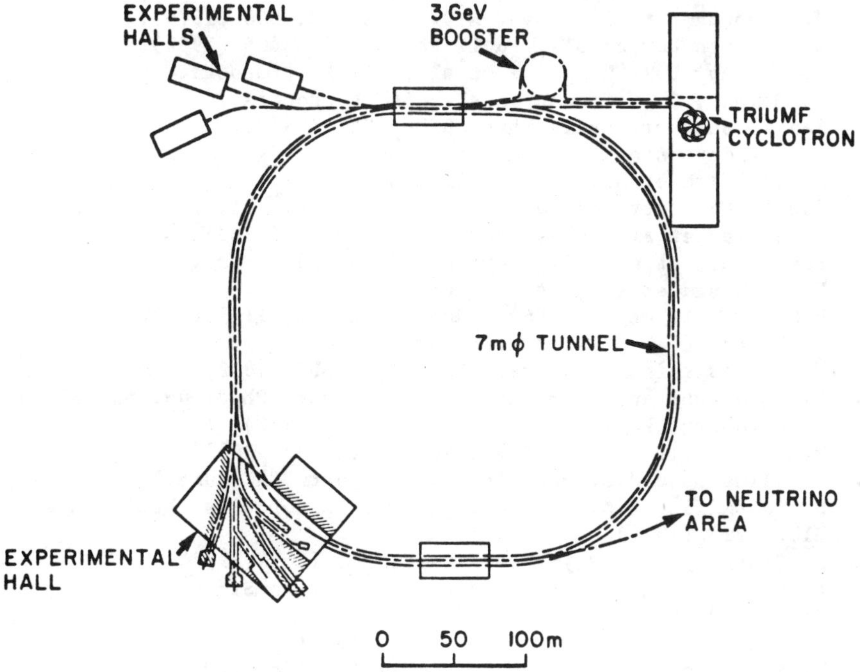

Fig. 1. Possible layout for a 30 GeV synchrotron booster for TRIUMF.

TABLE I Possible Synchrotron Specifications[25]

Final energy	16 GeV
Radius	80 m
Repetition rate	30 Hz
RF frequency	46 → 62 MHz
Intensity	3.3 μC/pulse
RF cavities	
Number	50
Power	250 kW
Length	2.39 m
Lattice	
Type	DFOFD (combined function)
Cells	30
Magnetic field max	11.36 kG
Tune	7.41
Transition γ	7.42 (6.03 GeV)
Injection technique	H^- stripping

REFERENCES

1. J.D. Bowman et al., Phys. Rev. Lett. 42, 556 (1979).
2. A. Badertscher et al., Nucl. Phys. A377, 406 (1982).
3. D. Bryman, C.K. Hargrove et al., TRIUMF proposal.
4. D. Bowman, C. Hoffmann et al., LAMPF proposal.
5. See O. Shanker, Nucl. Phys. B206, 253 (1982).
6. H.L. Anderson et al., LAMPF proposal;
   M. Strovink et al., TRIUMF proposal.
7. See R. Mohapatra and J.C. Pati, Phys. Rev. D11, 566 (1975).
8. D. Bryman et al., Phys. Rev. Lett. 50, 7 (1983).
9. See A. Bay et al., SIN Newsletter No. 14 (1982).
10. V. Highland et al., LAMPF proposal.
11. R.E. Mischke et al., Phys. Rev. Lett. 48, 1153 (1982).
12. A. Stetz et al., TRIUMF proposal.
13. L. Wolfenstein, Phys. Rev. Lett. 13, 562 (1962).
14. M. Kobayashi and T. Maskawa, Prog. Theor. Phys. 49, 652 (1973).
15. S. Weinberg, Phys. Rev. Lett. 37, 657 (1976).
16. M.K. Campbell et al., Phys. Rev. Lett. 47, 1032 (1981).
17. F. Gilman and J.S. Hagelin, SLAC report SLAC-PUB-3087 (1983).
18. See J. Ellis, M.K. Gaillard and D.V. Nanopoulos, Nucl. Phys. B109, 213 (1976).
19. D. Clark et al., Phys. Rev. Lett. 26, 1667 (1971).
20. R. Shrock, State University of New York at Stony Brook preprint ITP-SB-82-38 (1982).
21. Y. Asano et al., Phys. Lett. 107B, 159 (1981).
22. R. Shrock et al., Phys. Rev. Lett. 42, 1589 (1978).
23. B. Kayser et al., Phys. Rev. D20, 87 (1979).
24. M.K. Craddock, TRIUMF preprint TRI-PP-82-32.
25. L.C. Teng, "Some Details of the Kaon Factory", TRIUMF design note TRI-DN-82-2 (unpublished) 1982.

THE BNL AGS - A CONTEXT FOR KAON FACTORIES

L.S. Littenberg
Brookhaven National Laboratory
Upton, New York 11973

Figure 1 shows the Brookhaven site with the AGS-CBA complex highlighted. In this photograph the AGS is dwarfed by CBA and indeed during the past few years future plans for particle physics at BNL have been dominated by this enormous project. However, very recently interest in future physics use of the AGS has undergone a strong revival. Indeed, since the beginning of this year, two projects for augmenting the AGS have been proposed.[1,2] Such projects could keep the AGS viable as a research machine for many years to come. In general such schemes will also improve the performance and increase the versatility of the CBA, and so are doubly valuable.

It should be kept in mind that in spite of the fact the AGS has been perhaps the most fruitful machine in the history of high energy physics,[3] its full capacities have never been exploited. Even without improvements its usefulness as a source of new physics can be expected to continue for several years. At present it routinely produces 10^{13} protons/pulse in either slow extracted (one 1-sec spill every 2.4 seconds) or fast extracted (12 20-nsec bursts separated by 220 nsec every 1.5 seconds) mode. It is normally run at 28 GeV although its energy can be more or less continuously varied from 32 down to 1.5 GeV/c. In slow extracted mode the beam is shared by as many as 8 experiments and 2 test beams (see Figure 2). Table 1 lists the currently available beams. The capability for acceleration of polarized protons will soon be added.[4] Table 2 gives a summary of the experiments serviced over the last few years. Table 3 gives a breakup by subject of recent and proposed experiments. These experiments constitute quite a rich and varied program. I will discuss some of them individually a little later on. First, I have a couple of points to make about the economics of operating an accelerator in this energy range. Figure 3 shows physics output of the AGS (measured in experiment-weeks) versus operating cost in constant dollars. Note, that in recent years, this has been in the region of $30 M. This is a much higher figure than one often sees quoted as the operating cost of a kaon factory—which after all should be more expensive than the AGS to run. For $10 M one gets no physics at all, and there is a break in the slope at about $25 M. After this point one gets incrementally quite a bit more physics per dollar. Now although you and I think the sort of physics done at the AGS is exciting and important, someone influential has reservations because the most recent operating budgets have allowed $<$ 50% utilization of the machine. This is true in spite of the fact that we are operating on the steep part of the curve, where your dollar buys you most.

Next I want to discuss how well the AGS is doing at the kind of physics one would build a kaon factory to pursue. The area in which

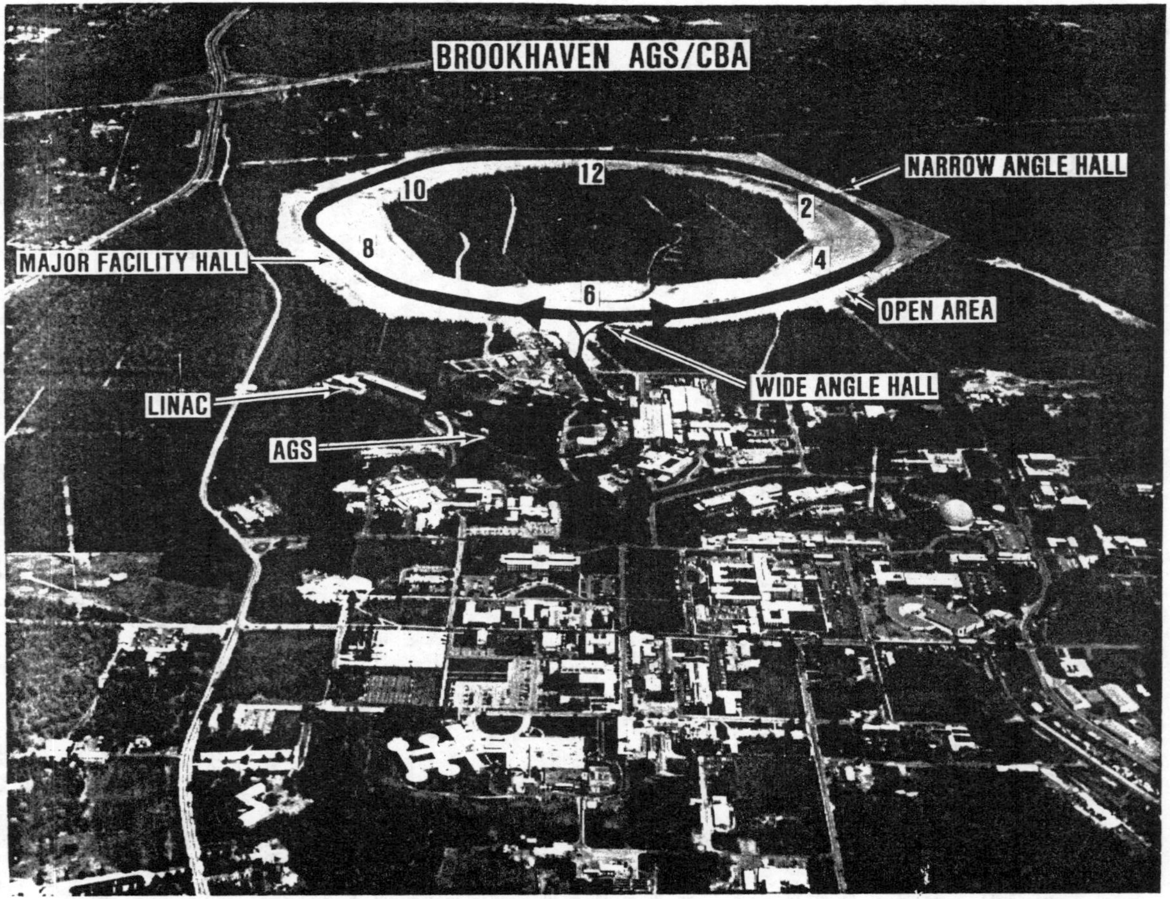

Fig. 1. Brookhaven site with AGS-CBA complex highlighted.

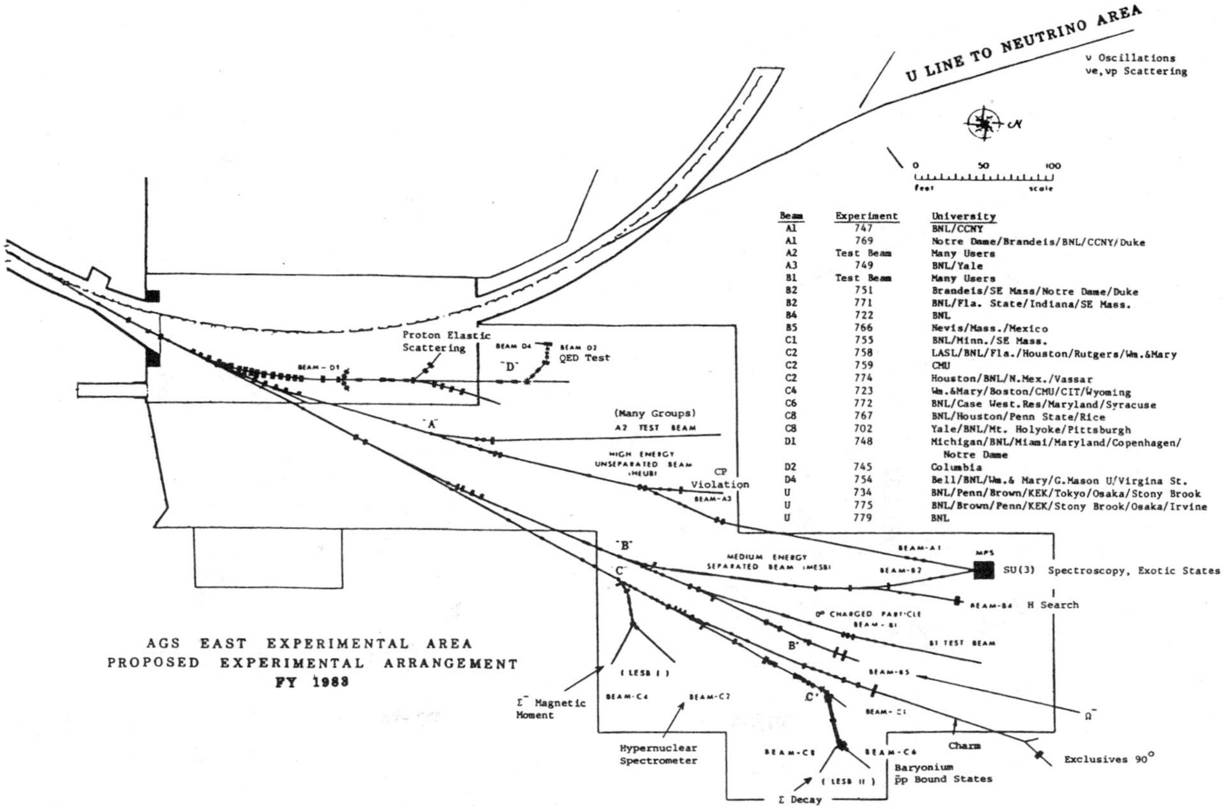

Beam	Experiment	University
A1	747	BNL/CCNY
A1	769	Notre Dame/Brandeis/BNL/CCNY/Duke
A2	Test Beam	Many Users
A3	749	BNL/Yale
B1	Test Beam	Many Users
B2	751	Brandeis/SE Mass/Notre Dame/Duke
B2	771	BNL/Fla. State/Indiana/SE Mass.
B4	722	BNL
B5	766	Nevis/Mass./Mexico
C1	755	BNL/Minn./SE Mass.
C2	758	LASL/BNL/Fla./Houston/Rutgers/Wm.&Mary
C2	759	CMU
C2	774	Houston/BNL/N.Mex./Vassar
C4	723	Wm.&Mary/Boston/CMU/CIT/Wyoming
C6	772	BNL/Case West.Res/Maryland/Syracuse
C8	767	BNL/Houston/Penn State/Rice
C8	702	Yale/BNL/Mt. Holyoke/Pittsburgh
D1	748	Michigan/BNL/Miami/Maryland/Copenhagen/ Notre Dame
D2	745	Columbia
D4	754	Bell/BNL/Wm.& Mary/G.Mason U/Virgina St.
U	734	BNL/Penn/Brown/KEK/Tokyo/Osaka/Stony Brook
U	775	BNL/Brown/Penn/KEK/Stony Brook/Osaka/Irvine
U	779	BNL

Fig. 2

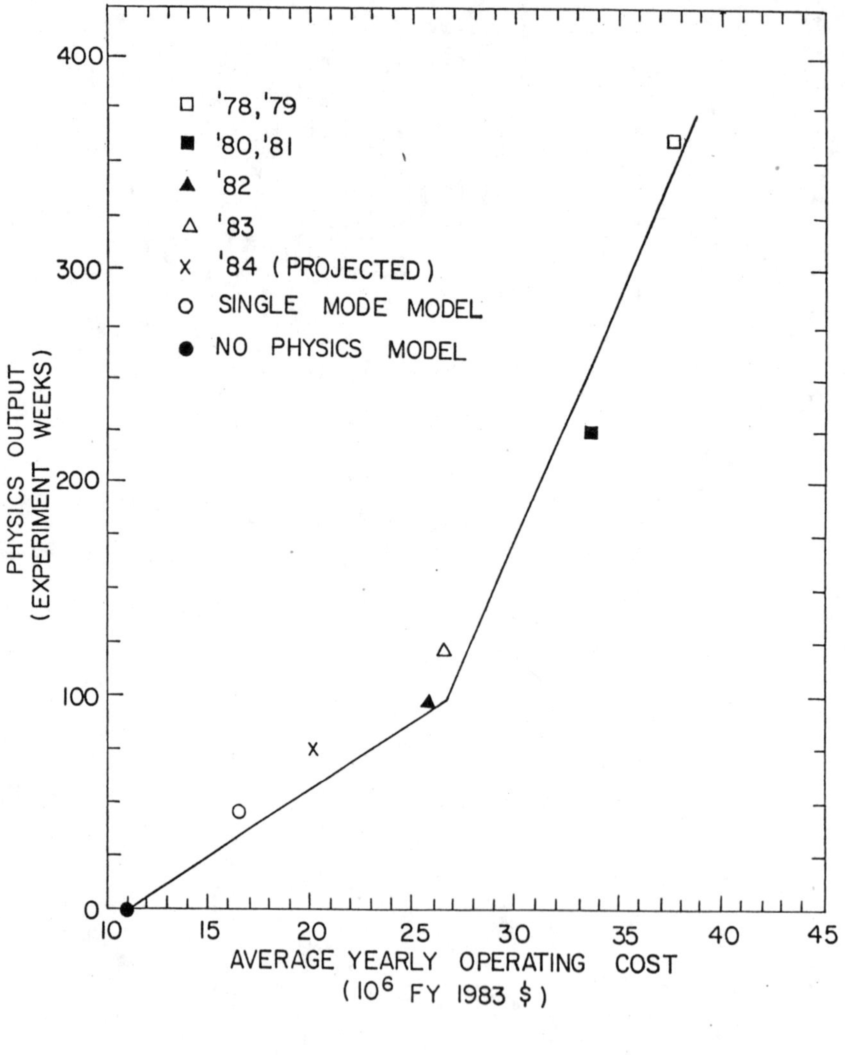

Fig. 3. Physics output in experiment weeks vs AGS operating budget in FY 1983 dollars.

BNL BEAMS OPERATIONAL
January 1983

Beam	GeV/c	±Δp/p (%)	Prod Angle	Ω (msr)	Flux in thousands/10^{12} protons on target: K^+	K^-	p	$\bar{p}$	π^+	π^-	@ GeV/c	Purity	Remarks
Separated Beams for General Use													
B4	1.5–6(K) 1.5–9(p)	3	3°	0.3	270	120	2×10^4	100	4×10^4	3×10^4	4	π^-/K^- 3 π/p 3/4	Usually 2×10^{12} ppp; L = 81 m.
C2,C4	$\leq$ 1.1	2	10.5°	2.6	40	12	2×10^4	2	8×10^4	8×10^4	0.75	π^-/K^- 10	Usually 2×10^{12} ppp; L = 15 m.
C6,C8	$\leq$ 0.8	2.5	5°	15.0	200	60	1.4×10^5	14	6×10^5	6×10^5	0.7	π^-/K^- 20	Usually 2×10^{12} ppp; L = 15 m.
Separated Beams for Fixed Facilities													
B2	(Same characteristics as B4)												MPS
Unseparated Charged Beams for General Use													
B1	5–24	3	0°	0.3	2500	700	1.5×10^5	200	6×10^4	3×10^4	10		Usually 2×10^{12} ppp; L = 75 m.
C1	5–24	5	0°	0.8	900	400	3×10^5	30	1×10^5	3×10^4	13	μ^-/π^-=.03	Usually 2×10^{12} ppp; L = 61 m.
Unseparated Charged Beams for Fixed Facilities													
A1	5–24	1.5	0°	0.2						1000	22		MPS: L = 130 m; 10^{12} ppp; 25 cm Be target
Neutral Beams for General Use								K_L	n				
A3	1–28		0°	0.0035				2000	10^5		1–28		$<10^{12}$ ppp; alternates with A1; L = 8 m; design intensity
B5	1–28		0°	0.01					3×10^5		1–28		10^{10} ppp typical; L = 2.6 m; design intensity
Muon Channel						μ^-							
D2,D4	1–.3(π) .05–.15(μ)	9(π)	55°(π)	50(π)		2000					0.10		Flux in 100 cm^2 with Δp/p = ± 2%; design intensity
Neutral Beam for General Purpose use						ν	$\bar{\nu}$						
U						$10^7/m^2$	$7\times10^6/m^2$					9×10^{12}ppp typical	Fast spill, flux avg. over 0.7 m radius, peaks at 1.5 GeV/c

*A3 has been converted to a neutral beam for FY 1982.

TABLE 1

TABLE 2

	FY80	FY81	FY82	FY83
AGS MANYEARS	359	314	278	270
AGS WEEKS	26	24	23	21
EXPERIMENTS SERVICED	10	9	13	17
NO. OF EXPERIMENTERS	181	142	168	251
NO. OF INSTITUTIONS	26	21	36	46

TABLE 3

EXPERIMENTAL TOPICS	
HADRON SPECTROSCOPY	13
HADRON DYNAMICS	7
SYMMETRY TESTS	7
DETAILED ELECTRO-WEAK INTS	6
HYPERNUCLEI	7
OTHER	3

the full capability of the AGS is most nearly exploited is that of ν physics. The entire internal beam is dumped on the ν target and the resulting horn-focussed decay beam impinges on a sophisticated 200 ton detector.[5] This consists of a series of alternating layers of liquid scintillator hodoscopes and proportional drift tube banks (see Fig. 4 and Table 4). Aiming primarily at the processes $\nu_\mu + e^- \rightarrow e^- + \nu_\mu$ and $\bar{\nu}_\mu + e^- \rightarrow e^- + \bar{\nu}_\mu$ its designers took pains to keep the fraction of inactive material in the detector to an absolute minimum. In this way they hope to be able to distinguish electron showers from the far more copious γ showers resulting from processes like $\nu_\mu + A \rightarrow \nu_\mu + \pi^0 + X$. How well this worked is illustrated dramatically by Figure 5 which gives the θ^2 distribution of events identified as electrons versus those identified as photons. There is a clear peak at $\theta^2 \approx 0$ in the 'electron' events as expected for νe elastic scatters, whereas the 'photon' events have no such structure.

Next we move on to a pair of recently approved K decay proposals. I will be brief about these since proponents of each of them are present at the conference and are available to discuss them in detail. Both experiments search for possible lepton flavor violations: E780[6] will probe $K^0{}_L \rightarrow \mu e$ to $< 10^{-10}$ and E777[7] will probe $K^+ \rightarrow \pi^+ e^- \mu^+$ to 10^{-11}. The E780 apparatus is shown schematically in Figure 6. A beam of 2×10^7 $K^0{}_L$ (and 6×10^8 neutrons) impinges on a drift chamber spectrometer. About 1% of the $K^0{}_L$ decay in the drift zone and about 10% of the two-body decays are accepted by the spectrometer. In a 1000 hr experiment this allows them to get beyond 10^{-10} in statistical sensitivity. Of course, to justify this, the background rejection must be commensurate. Electrons are identified by a segmented hydrogen Cerenkov counter and a lead glass hodoscope. Muons are distinguished from π's by means of a hadron filter. In this experiment particle identification is of limited value because the main background arises from $K^0{}_L \rightarrow \pi e \nu$ decay in which the π subsequently decays to a μ somewhere in the apparatus. Although the apparatus has been made as short as possible to minimize this, once such a decay has occurred the only recourse is to kinematics. Therefore, great stress is placed upon resolution. The $K^0{}_L \rightarrow \mu e$ is generally a 3-c fit since one knows the incoming $K^0{}_L$ direction. In addition there is a vertex constraint and consistency checks given by redundant chambers. Since one is shooting for a unique point in the parameter space given by the kinematic, vertex, and other constraints, whereas the background is smoothly distributed in this space, the better the resolution, the smaller the target region and consequently the smaller the contamination. Crudely one tends to gain as a power $n > 2$ of the spatial resolution of the chambers. However, at a certain point, multiple Coulomb scattering in the chambers make further gains in resolution almost impossible. This point seems to be reached at a contamination level of about 10^{-11} according to the proponents of E780.[8] As their spectrometer will be operating with very nearly this optimal

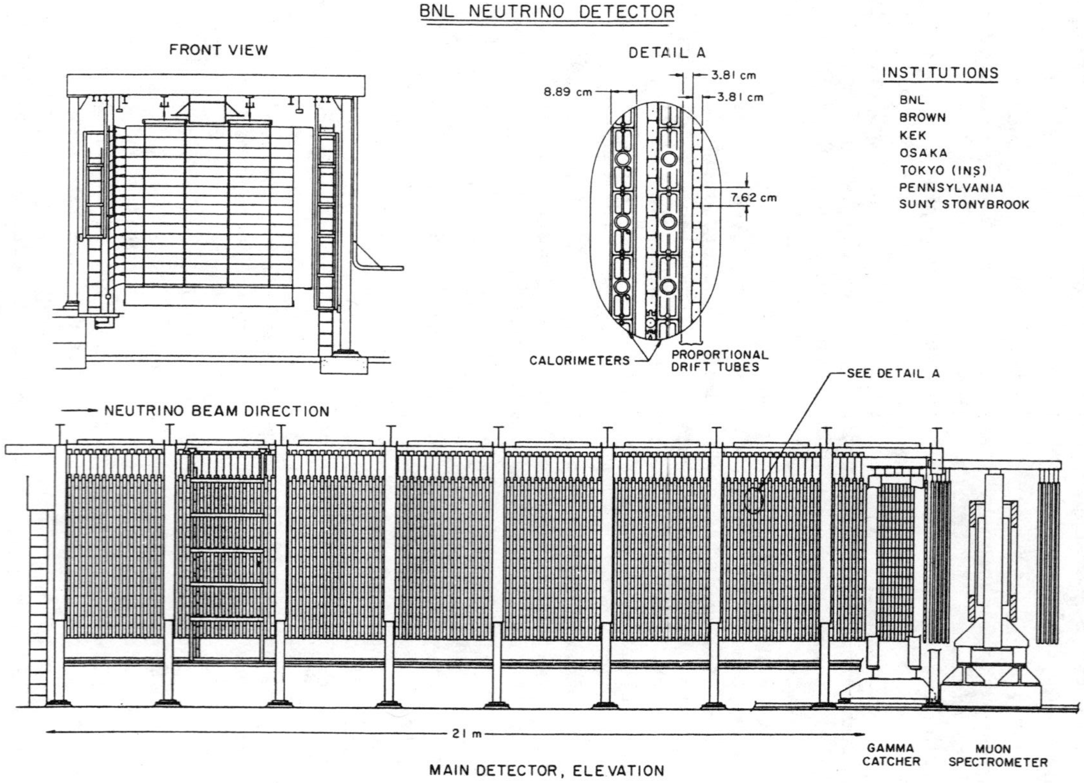

Fig. 4. The BNL ν detector.

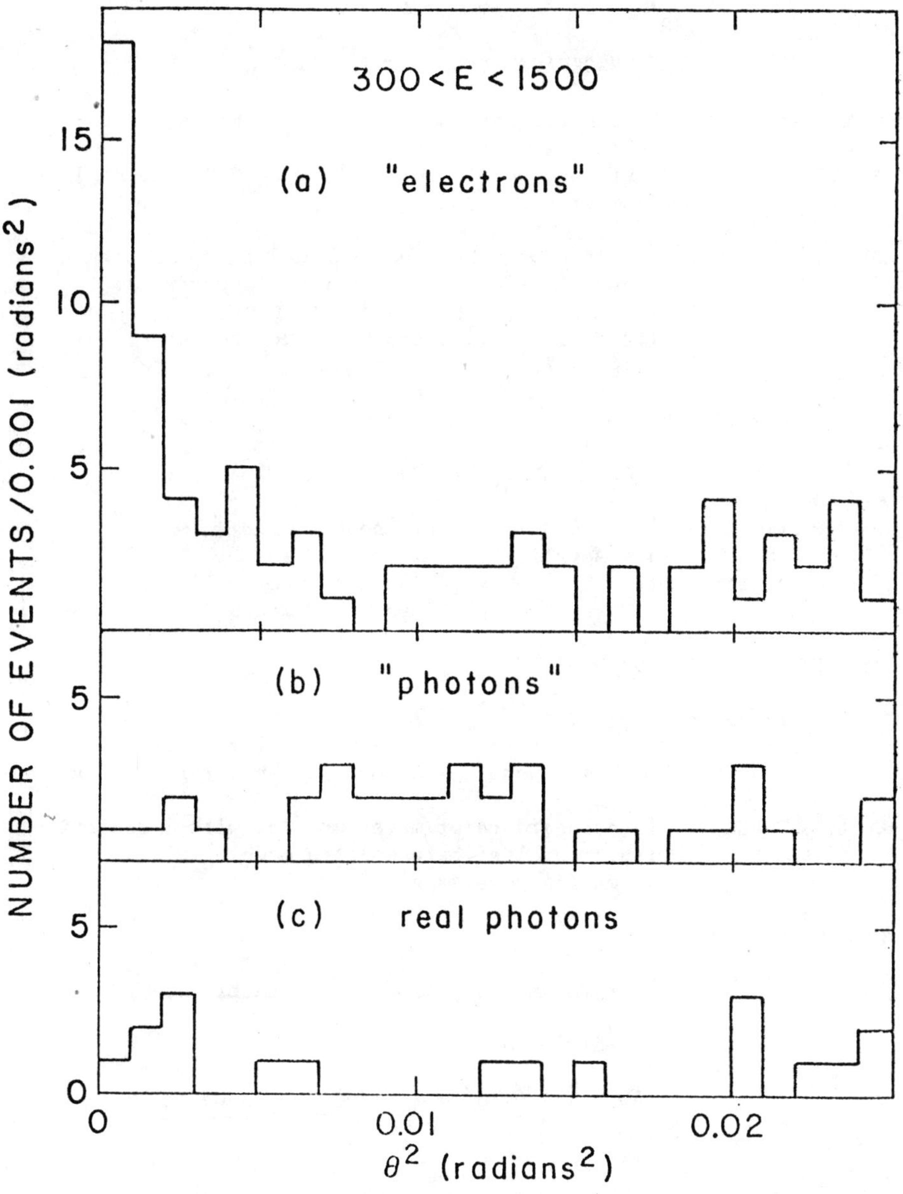

Fig. 5. Preliminary data from Experiment 734. Distributions in the quantity θ^2 for single shower events: a) Events in which the shower was identified as an electron, b) events in which the shower was identified as a photon, c) control sample photon events.

TABLE 4

BROOKHAVEN NEUTRINO DETECTOR

LOCATION	Brookhaven National Laboratory, Upton, NY 11973
INCIDENT BEAM	Neutrino, Horn Focussed from 28 GeV Protons 110m from proton target
ASSEMBLY	Modular construction; each module consisting of a plane of calorimeter and two planes (x,y) of tracking proportional drift tubes 112 Modules + γ-Catcher + Spectrometer Weight 172 + 30 metric tons

MODULE PROPERTIES

CALORIMETER (LIQUID SCINTILLATOR)

Active Area 4.22 × 4.09 m^2 Thickness 7.9 cm
Weight (Liquid & Acrylic) 1.35 metric tons
16 cells/module 2 Amperex 2212A phototubes/cell
1 Pules Height Measurement/2 Time Measurements Per Tube Readout

PROPORTIONAL TUBES (PDT)

Active Area 4.2 × 4.2 m^2
Thickness (x and y) 7.6 cm 54 x wires 54 y wires
1 Pulse Height Measurement/2 Time Measurements Per Wire Readout

GAMMA CATCHER	10 standard calorimeter modules with 1 radiation length of lead between each module 30 metric tons target mass.
MUON SPECTROMETER	
	2m × 2m Aperture Muon Spectrometer
	$\langle \int Bdl \rangle = 70$ MeV/c
	$(\Delta p/p)^2 = (.10^2 + (.067p)^2)$ p in GeV/c

resolution they expect to reach the 10^{-10} level fairly easily. The experiment is also sensitive to $K^0_L \to \mu^+\mu^-$ and e^+e^- at about this level.

The E777 apparatus is shown in Figure 7. An unseparated 6 GeV/c beam is incident on a PWC spectrometer. There are again about 20 million K^+, this time accompanied by 2.5×10^8 π^+, and 1.2×10^8 protons per spill. At the front end of the spectrometer is a pitching magnet to bend the K^+ decay products out of the intense beam and to separate them by charge. On the negative track side electrons are identified and heavy particles rejected by two H_2 Cerenkov counters and a lead glass hodoscope. On the positive side electrons will be rejected by two CO_2 or N_2 Cerenkov counters and more lead glass, whereas muons will be distinguished from pions via a hadron filter. This heavy artillery is motivated by the fact that in this experiment, unlike $K^0_L \to \mu e$, a large background suppression through particle identification is possible. However, once again the final recourse is to kinematics so that good resolution is also essential. About 10% of the K^+ decay in the drift region, and the acceptance for $K^+ \to \pi^+ e^- \mu^+$ is about 10%. This yields an overall statistical sensitivity of 1 per 2×10^{11} in 10^6 pulses. Residual backgrounds are estimated to be at the few $\times 10^{-12}$ level, making a sensitivity of $\sim 10^{-11}$ feasible. The experiment may also be able to look at $K^+ \to \pi^+ e^+ e^-$, $\pi^0 \to e^+e^-$ and other interesting processes.

A third interesting K decay mode is $K^+ \to \pi^+ \nu\bar{\nu}$. Although no proposal has been submitted, an experiment sensitive to this mode at the 10^{-9} level is being planned. A possible detector is sketched in Figure 8. One would attempt to stop about 10^5 K^+/pulse in a highly segmented 20 cm target. The full beam rate through this target would then be roughly a megacycle (mainly π^+). Charged decay products are momentum analyzed in a cylindrical drift chamber system inside a superconducting solenoid. Range and energy of the daughters are measured in a cylindrical scintillator array. Thus pions can be distinguished from muons by comparing range and momentum. In addition one detects the π-μ-e decay chain in the scintillator array. The main backgrounds come from the dominant $K^+ \to \mu\nu$ and $\pi^+\pi^0$ modes and from $K^+ \to \mu\nu\gamma$. Photon vetos surround the entire apparatus in an effort to suppress the latter two backgrounds. Since there is no unique kinematic signature for $K^+ \to \pi^+\nu\bar{\nu}$, background rejection is absolutely crucial. Two body decays such as $K^+ \to \pi^+$ + Familon as proposed by Wilczek[9] are quite a bit easier to detect. Although the apparatus as shown has a geometrical acceptance of ~ 50%, and accepts 50% of the π^+ momentum spectrum, the overall detection efficiency is only ~ 2%. This is due to a number of individually small or moderate losses (e.g., pions interacting before ranging out, pulse height and timing cuts on decaying particles etc.). Thus for 10^6 pulses at 10^5 K^+ stops/pulse one gets to the 10^{-9} level. One could also study $K^+ \to \pi^+\tilde{\gamma}\tilde{\gamma}$, $K^+ \to \mu\nu_{heavy}$, etc., with this apparatus.

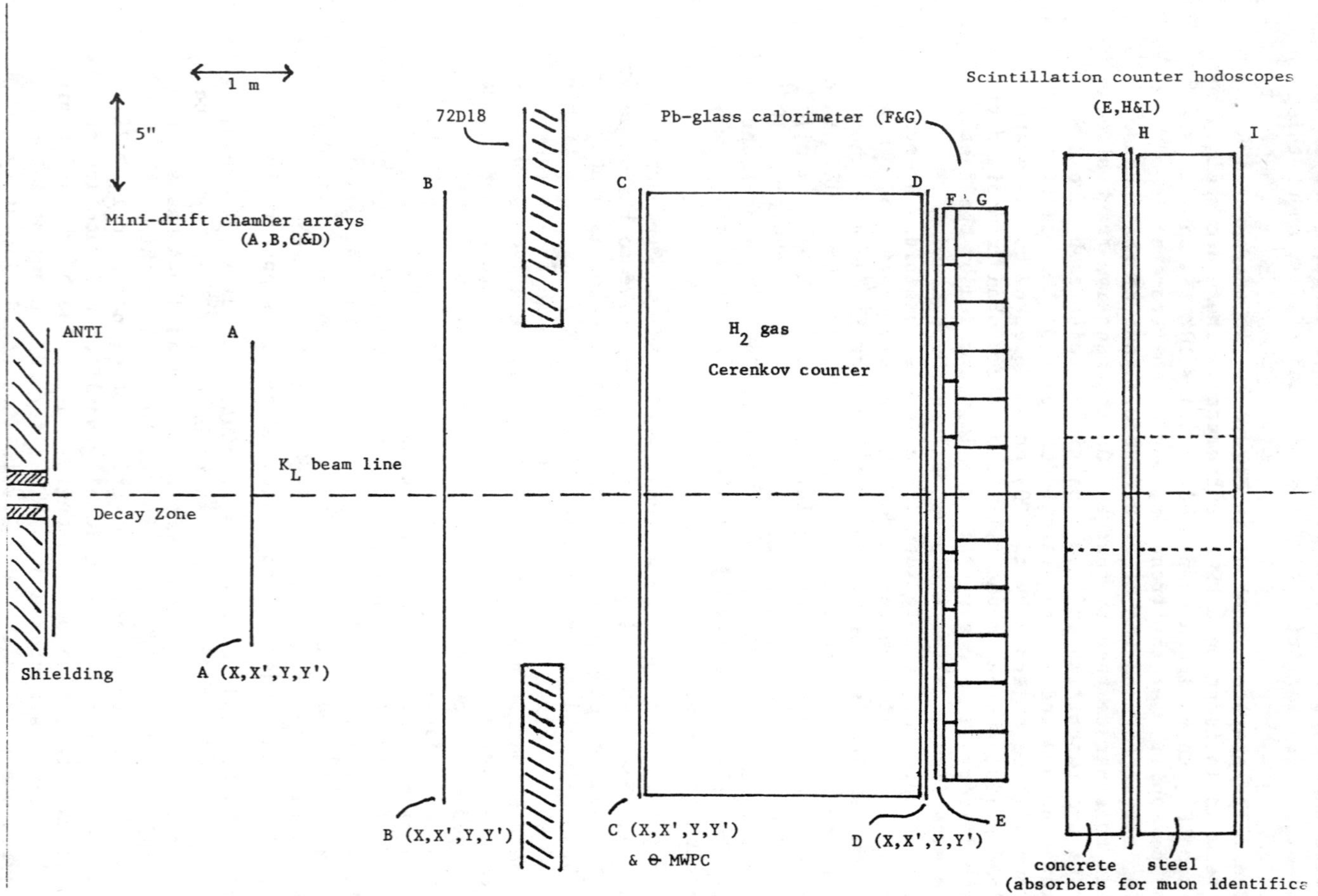

Fig. 6. Schematic of E780 apparatus.

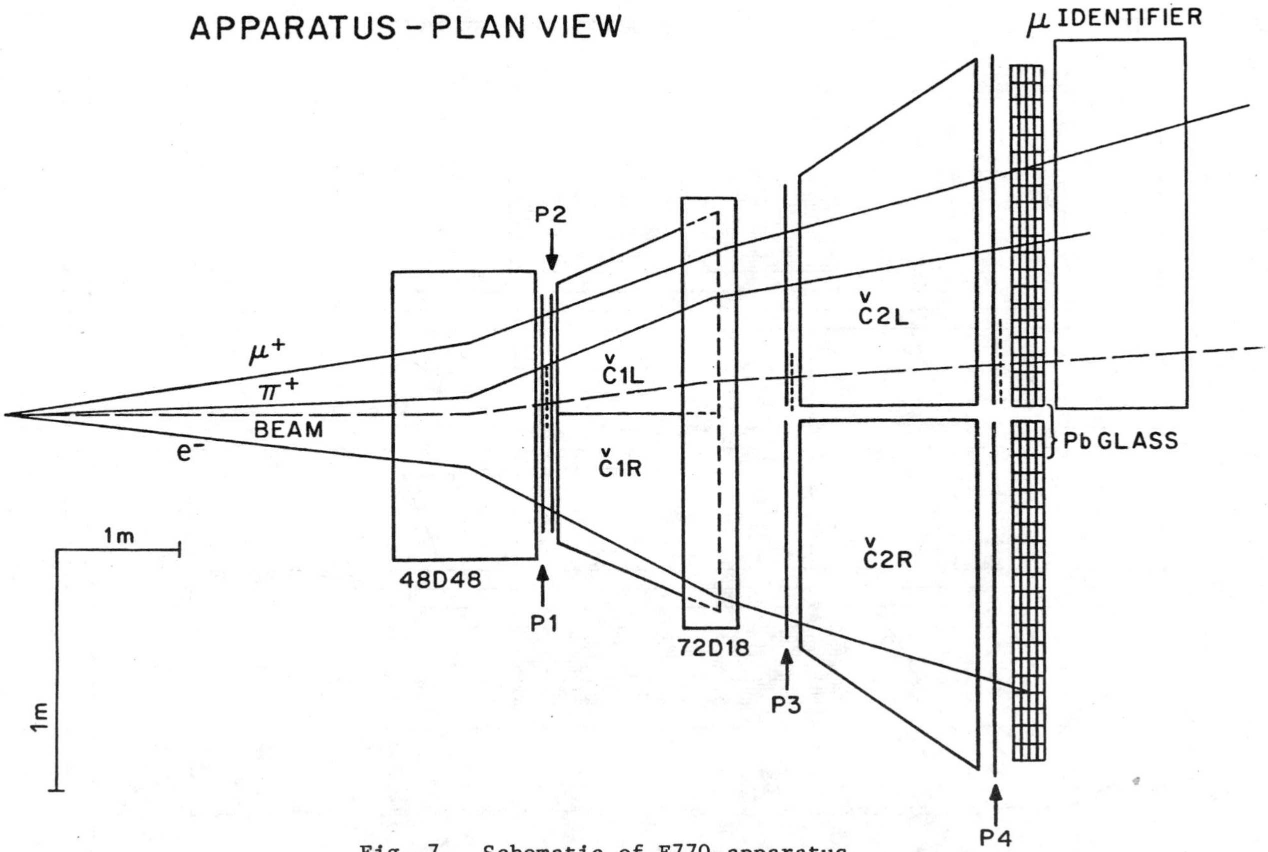

Fig. 7. Schematic of E770 apparatus.

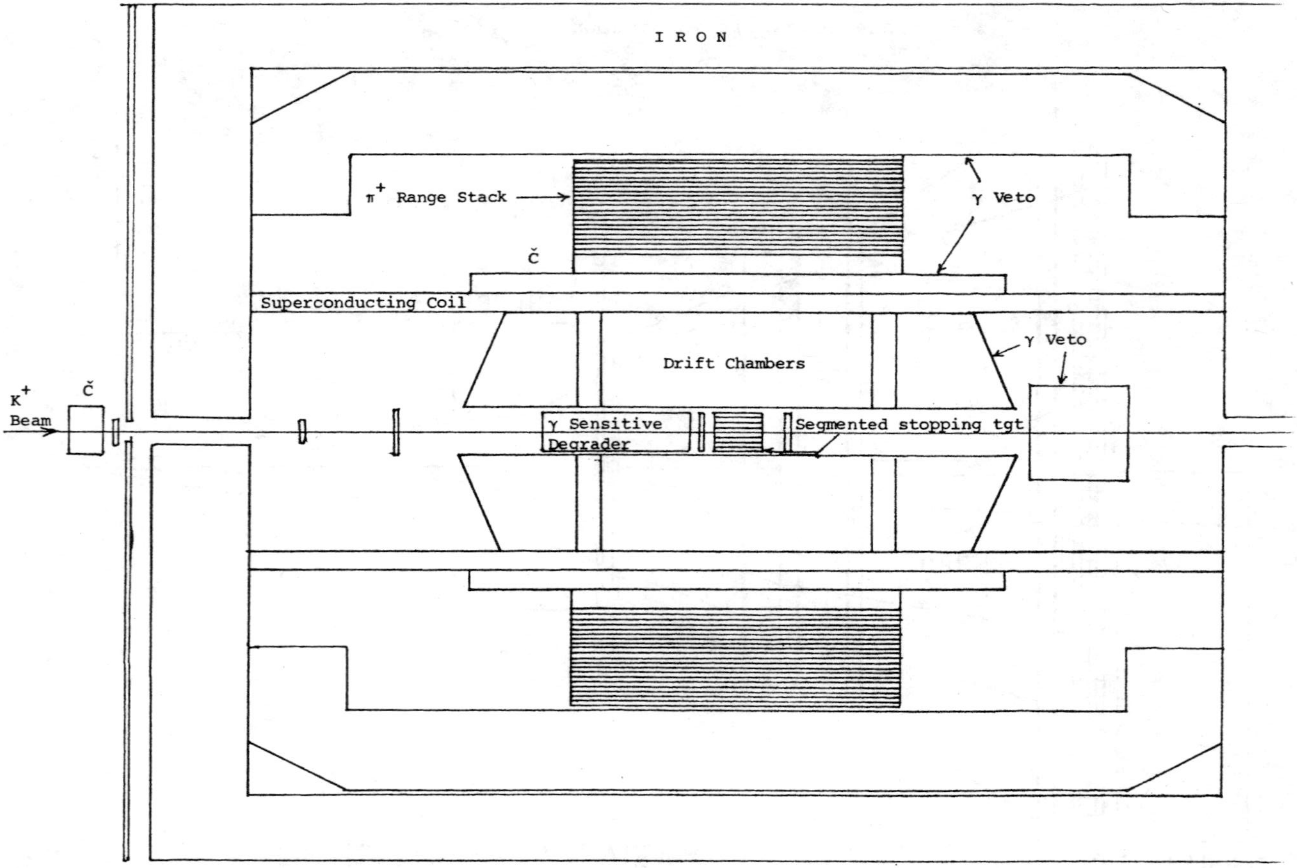

Fig. 8. Possible detector for $K^+ \rightarrow \pi^+ \nu\bar{\nu}$.

Now I'd like to take up the subject of how far these sorts of experiments might be taken at the AGS. It seems hard to increase the number of ν_μ events without increasing the primary beam intensity. In principle, with enough money and no competition from the slow beam program, one could run four or five times longer than has been the case in recent years. One could also think about making the detector larger (or adding more detectors). However, the present detector cost \$6 M and its builders feel that it is already large enough to have realized most of the possible economies of scale. Thus, to price a larger detector of comparable quality just multiply \$6 M by the ratio of tonnage. Of course, special purpose detectors of more limited usefulness could cost less per ton (one has already been approved for a ν oscillation experiment[10]). It's somewhat daunting to note that the present beam-apparatus combination is not all that far from being rate limited (for ν_μ). Although the experimenters make very good use of the time structure of the beam to reduce background (see Figure 9), the in-time rate approaches 1/spill. Since the apparatus is constantly sensitive (triggerless) it tends to get confused by more than a few events/spill. Thus one can imagine using perhaps another order of magnitude but not much more in such an apparatus. By contrast, almost unlimited increases in flux would be welcome to study ν_e and $\bar{\nu}_e$ interactions, or to be traded in for ν energy determination by making a narrow band beam.

As mentioned above E780 is not expected to be background limited. Thus if the mini-drift chambers are not overloaded, such an experiment could use more beam. Unfortunately, because of the dearth of experience with 0^0 $K^0{}_L$ beams, one doesn't yet know the ultimate $K^0{}_L$ intensity and purity obtainable at the AGS. In fact, the proponents of E780 are now carrying out a CP-violation experiment[11] in such a beam and will soon settle the issue. In the meantime one must rely on calculation and extrapolation from other data. This turns out to be rather difficult because data near 30 GeV is sparse and of fair quality at best. Data near 200 GeV is better but the extrapolation is a long one. Secondly, most experimental results are off H_2 and so need to be corrected for A dependence in order to predict the yield of a production target. There are also corrections for multiple interactions, beam attenuation etc., which are not entirely cut and dried. Two recent attempts[6,12] to calculate the flux disagree by a factor 2 or 3. If the more optimistic estimate[5] is right, the n/K ratio in the beam will be around 30; if the more pessimistic is correct, it will be more like 100. Figure 10 shows the flux vs production angle for the pessimistic calculation. Using these conservative numbers, one could get 6.5×10^5 usable $K^0{}_L/10^{12}$p/μsr. As much as 4×10^{12} ppp have been delivered to targets at the AGS in the past, and solid angles up to ~ 250 μsr have been used. (E780 will use $\sim 10^{12}$ protons and $\Delta\Omega \approx 15$ μsr). This would yield at least 6.5×10^8 $K^0{}_L$ pulse. For a decay tank of length .03 Λ_K and an apparatus of

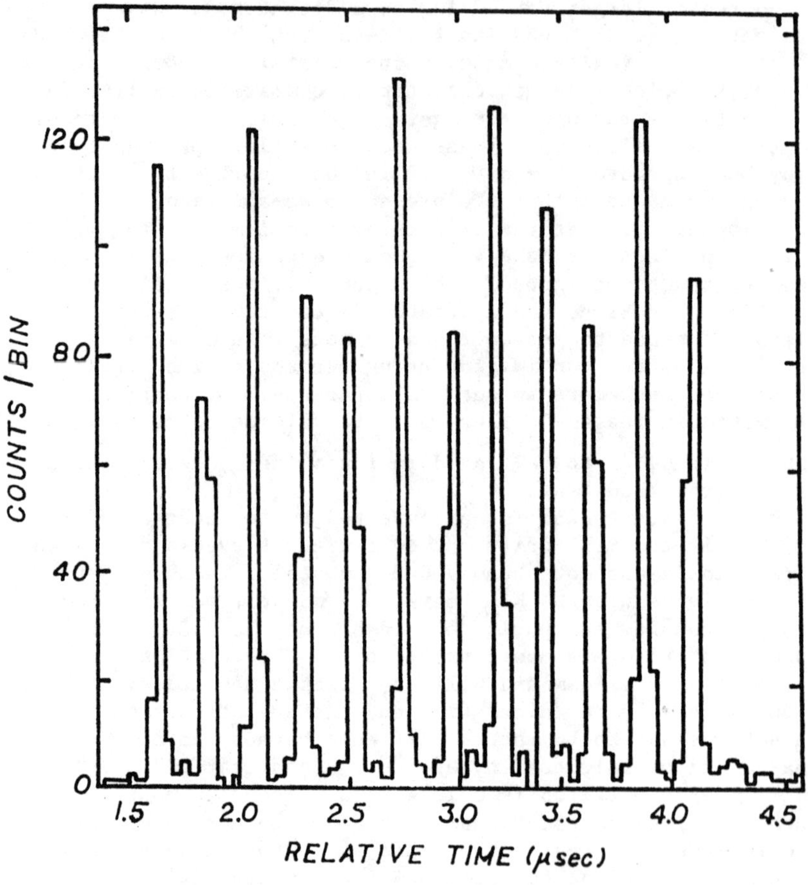

Fig. 9. Time distribution of ν events from E734.

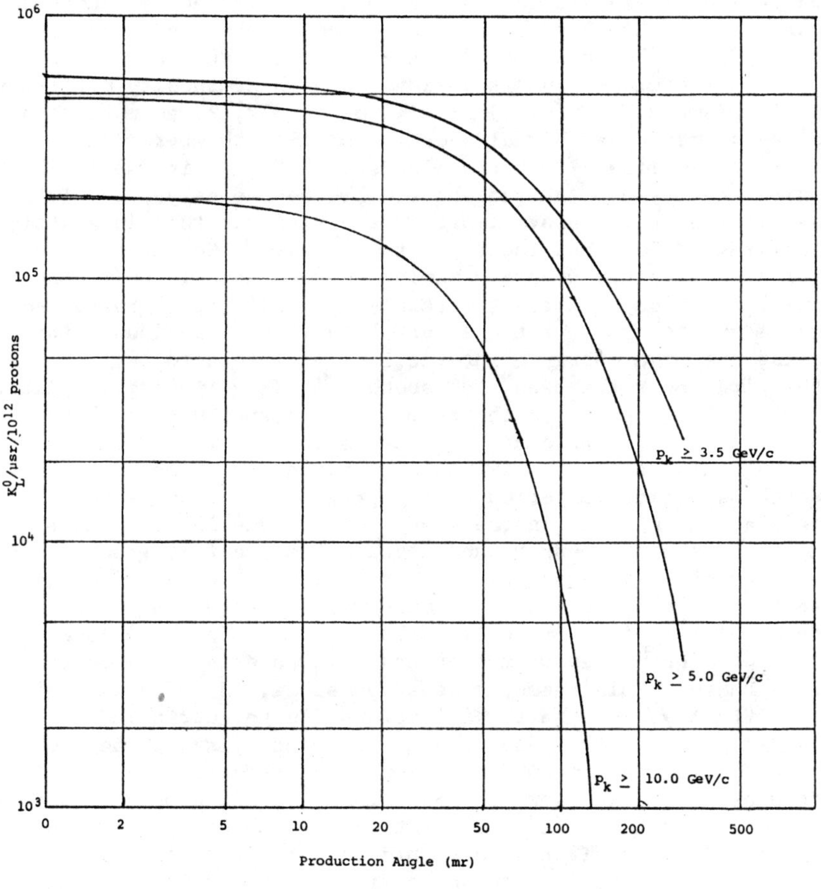

Fig. 10. K^0_L flux per 10^{12} protons on target per μsr vs production angle for various p_K ranges (from Ref. 12). Assumes 28.5 GeV/c protons and one interaction length Be target.

5% acceptance, one would reach a statistical sensitivity of ~ 10^{12} in a 1000 hr experiment. The optimistic estimate of K^0_L flux would yield ~ 3×10^{-13}! The K^+ situation is similar. Here we have a measured number of about 2×10^5 $K^+/10^{12}$p/μsr at 4 GeV/c with $\Delta p/p$ ~ 10%. For $\Delta\Omega$ = 500 μsr and 4×10^{12} on target one would get 4×10^8 K^+/pulse accompanied by 20 times more p's and π's. Then the statistical sensitivity attainable is similar to that of the K^0 case. Now these are formidable sensitivities and we must ask what barriers, other than raw K flux, stand in the way of achieving them. First the question of rates in the detectors. If we adopt the criteria developed in discussions of high luminosity detectors for CBA[13] we must satisfy: 1) rate/wire < 2 MHz; 2) no more than 1 or 2 pile-up events per signal event. For a 2m chamber with 2mm wire spacing the first limit translates to 2×10^9 hits/sec! The K^0_L's in the hot beam discussed above will provide only ~ 3×10^7. Thus one is within the above limit if the hits due to n interactions are less than ~ 30 times those due to K decays. Now this may seem quite safe, since we are assuming no more than 100 n/K, and only a small fraction of these (certainly not 1/3!) can be expected to cause chamber hits, but one must be a little cautious. For one thing the high energy n interactions will tend to be concentrated in the wires near the beam, for another it is quite common in actual fact to have ten times the chamber rates that one calculates in advance. As for pile-up, with a drift time of say 50 nsec, the average K decay event will be accompanied by .6 accidental event which satisfies criterion 2. In principle, triggers can also be made at these rates in the sense that triggering devices e.g., multi-element C counters and electronics are fast enough, but the problem of making the trigger for e.g., $K \to \mu e$ sufficiently selective is non-trivial. Events with an e and apparent μ from K_{e3} decay will impinge on the detector at a rate of > 10^5/sec. Kinematic qualities of the events will have to be exploited at rates like these. This seems at least possible, if not easy.

Finally, we come to limitations due to intrinsic physics background. In the case of $K^0_L \to \mu e$, according to the proponents of E780, one can improve the background rejection by reducing the chamber resolution down to about 150μ, at which point MCS in the planes dominates. At this point one has about one background event/10^{11} sensitivity. Now this is not the limit of course. With enough statistics one can see a signal above quite a large background. However, to be generally accepted, a history making result, such as the observation of $K^0_L \to e\mu$, would be required to be of very high quality, certainly a > 5σ effect. Assuming the above background, even with a statistical sensitivity of 3×10^{-13} then, one could not claim a signal much less than ~ 28 events (= 8.5×10^{-12}) i.e., about the background level. If one had an order of magnitude more K's, this limit would only go down to 2.7×10^{-12}. This illustrates the general point that once a background is present, one tends to gain only as the square-root of the increase in statistical sensitivity.

The situation is similar for $K^+ \to \pi^+ e^- \mu^+$. Assuming 10% of decays occur within the fiducial region and a detection efficiency for these of 10%, in 1000 hours one could reach a statistical sensitivity level of 2.5×10^{-13}. In E777 the proposed sensitivity is limited by a desire to keep the chamber rates from $K^+ \to \mu^+ \nu$ decays below a few megacycles. Compared with what we have discussed above this is very conservative indeed. In spite of the fact that the PWC resolution is nowhere near the state of the art, the expected residual background is at the few $\times 10^{-12}$ level. Thus it does not seem impossible that the background could ultimately be brought down to the 2.5×10^{-13} level of the statistical sensitivity. This would allow one to pull out a convincing signal at the $\sim 10^{-12}$ level. I should inject a note of realism here and point out that both these ultimate experiments would be massive undertakings involving techniques that are somewhat beyond the present state of the art. It would take several years of unrelenting effort to reach the target sensitivity levels, even if no anticipated sources of background or other impediments were to show up. Elsewhere[14] I have discussed at length the sort of unexpected problems that can plague rare K decay experiments.

Thus both $K^0_L \to \mu e$ and $K^+ \to \pi^+ e^- \mu^+$ experiments could probably use the maximum flux available at the AGS. This may not be true of $K^+ \to \pi^+ \nu\bar{\nu}$ which is more likely to be background limited. A recent test in the LESB I[15] has shown that at $p_{K^+} = 710$ MeV/c one can stop well over 200,000 $K^+/4 \times 10^{12}$ protons on target. This number of protons/pulse has been routinely delivered to the C target for some time now. The π/K ratio in the beam was $\sim$ 3-4 to 1 and about 40% of the beam K^+ were stopped in a lucite target 10 cm in diameter and 20 cm long. Raising the beam momentum to 750 MeV/c increased the yield of stopped K^+/proton on target by a factor 1.7 without significantly degrading the transverse stopping distribution. The time available for the test ran out before the $p_{K^+} = 800$ MeV/c setting could be optimized, but it was clear that another large increment in stopping K^+/proton on target could be realized. At about this point one would probably start compromising beam properties (longer and wider stopping target necessary, more beam π^+/stopping K^+ etc.). However, it seems clear that one could obtain upwards of 500,000 K^+ per pulse under reasonable beam conditions. Presumably one could stop even more K^+ in the LESB II,[16] albeit under worse conditions. Since the present BNL beams are surely not the ultimate one could design, it does not seem fantastical that one could eventually stop $\sim$ 1 M K^+/pulse in good conditions at the AGS.

Now the question is whether one could begin to use such a flux in an experiment like $K^+ \to \pi^+ \nu\bar{\nu}$. Since at least one detector for this reaction with a 20% acceptance has been proposed[17] at CERN, one can imagine statistical sensitivities in the 10^{-11} ballpark. But will the backgrounds permit this level to be realized? Take the decay $K^+ \to \mu^+ \nu \gamma$ which occurs at about 3×10^{-3} in the kinematic region of interest. One would need to reject this by at least 10^{-9} to reach the sensitivity for $K^+ \to \pi^+ \nu\bar{\nu}$ discussed above. One can probably get $> 10^{-3}$ rejection of μ's by comparing energy and range. The experiment[18] which established the present limit of $BR(K^+ \to \pi^+ \nu\bar{\nu}) < 1.4 \times 10^{-7}$ used the observation of the $\pi \to \mu \to e$ decay chain to reject μ's by about 10^5 to 1. How well this technique works depends entirely on the accidental rate in the π range stack. Since the beam rate in the previous experiment was ~ 300 KHz, while that of a 1 MHz stopping K^+ beam would be more like 15 MHz, to get an equally good result one would need to increase the detector granularity by a factor 50. This might well be prohibitively expensive,[19] and still might not work--it was valuable in the previous experiment to veto any apparent decays which occurred within about 50 nsec of a subsequent beam track. Since the particles in a 15 MHz beam come every 70 nsec or so, it seems impossible in the case being considered here. Thus I'd guess one would be lucky to get a μ rejection factor of 10^3 by waiting for the π-μ decay only. One then needs something like a 10^3 rejection of photons to reach the necessary suppression of $K^+ \to \mu^+ \nu\gamma$. Since at low energy the photon attenuation length is about twice the radiation length, about 14 radiation lengths of veto detector are necessary to achieve this. Even then, any cracks, dead spaces or inactive absorber undermine this rejection. What's more, at the beam rate under discussion, random vetoing is liable to induce unacceptable deadtime unless the veto device is extremely fast. Other backgrounds produce equally daunting problems at these rates. Thus 10^{-11} seems like the hairy edge for this process at the AGS.

Now although I concluded that each of the K decay experiments considered above could be pursued at the AGS to about the point that intrinsic backgrounds allow, each would be a massive undertaking, tending to dominate the AGS program. It's hard to see how all three plus the other interesting physics of the energy range could be pursued simultaneously. To alleviate these problems one may rightly ask whether the AGS intensity could be increased.

This brings us to the subject of AGS future plans. As mentioned above a program allowing the acceleration of polarized protons is under way at the AGS. The initial intensity is anticipated to be around 10^{10} protons/pulse. Although this is adequate for most polarized proton experiments, such an intensity would, in general, be useless to unpolarized beam experiments. In addition the most interesting uses of polarized protons in CBA require the full luminosity. Thus to avoid program conflicts it would be desirable eventually to raise the polarized proton intensity to 10^{13} protons/pulse. To accomplish this one probably needs both a greatly improved ion source and an accumulator ring. A proposal[2] to build such a ring has recently appeared. Figure 11 shows the layout of the ring which would be situated between the present 200 MeV LINAC and the old 50 MeV LINAC building. The parameters of the ring are given in Table 5. Twenty 200 MeV LINAC

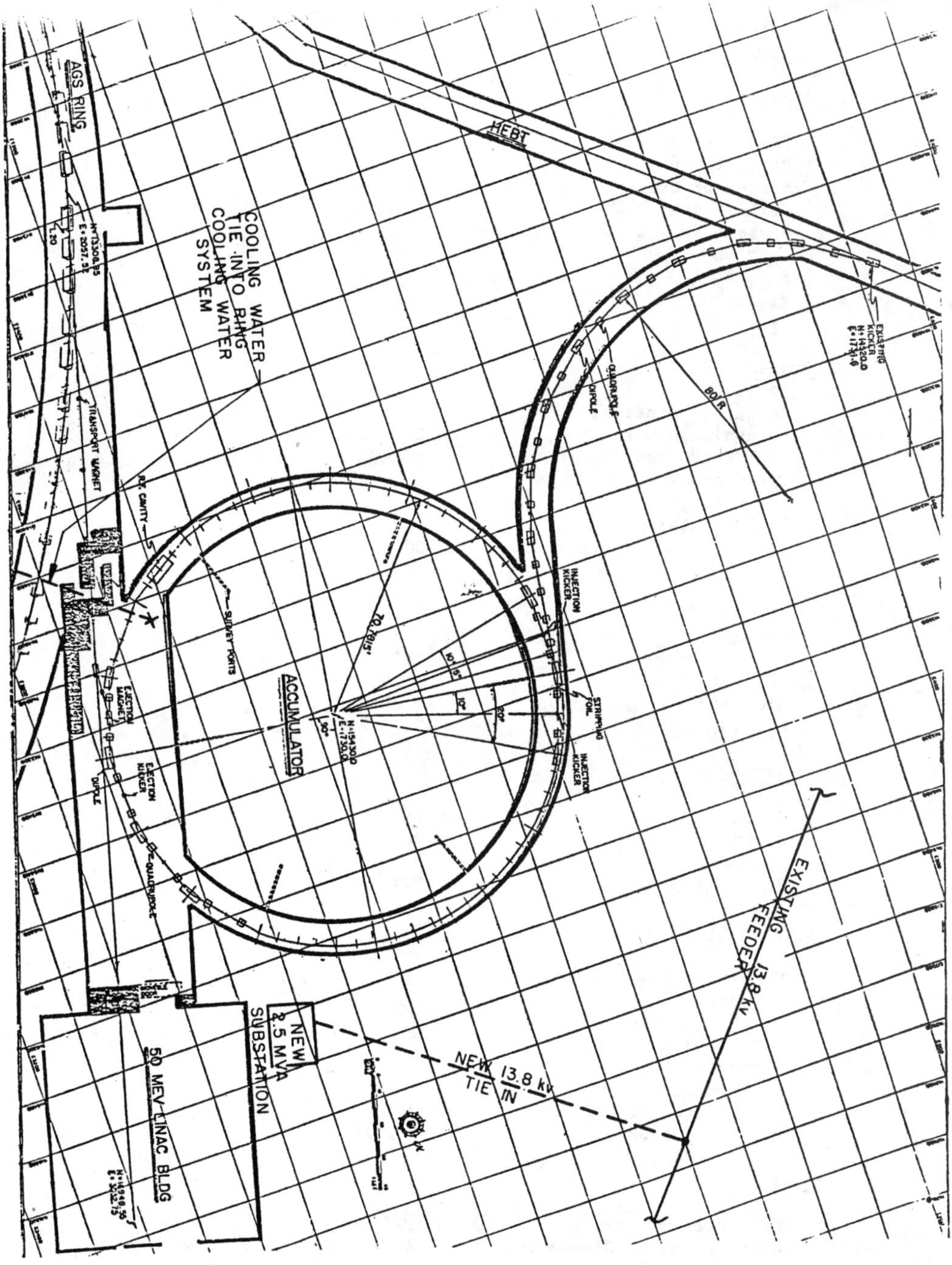
AGS RING
HEBT
COOLING WATER
TIE INTO RING
COOLING WATER
SYSTEM
TRANSPORT MAGNET
R.F. CAVITY
SURVEY PORTS
ACCUMULATOR
INJECTION KICKER
STRIPPING FOIL
INJECTION KICKER
EJECTION MAGNET
EJECTION KICKER
DIPOLE
QUADRUPOLE
EXISTING KICKER
EXISTING
FEEDER 13.8 kv
NEW 13.8 kv
TIE IN
NEW
2.5 MVA
SUBSTATION
50 MEV LINAC BLDG

TABLE 5

Accumulator/Booster Parameters

Energy

Injection	200 MeV
Maximum	1000 MeV

Lattice

Circumference	134.52 m (1/6 AGS)
Periodicity	18
Number of Cells	18 FODO
Cell Length	7.473 m
Phase Adv/Cell	94°
$\nu_x - \nu_y$	4.7
β_{max}/β_{min}	12/2 m
η_{max}	2 m
γ_T	4.55

RF System

Number of Stations	1
Harmonic Number	2
Peak RF Voltage	18 kV
Synchronous Phase Angle	
Capture	0°
Transfer	5°
Acceleration (future)	30°

Dipole

Number	18
Length (iron)	75"/78.25" (199 cm)
Gap	3.25"
Good Field Region	> 7"
Field (Inj/Max.)	3.73 kG/9.82 kG

Quadrupole

Number	36
Length (iron)	20"
Aperture	8" ϕ
Pole Tip Field	3.75 kG

Injection Line

Number of Dipoles (angle)	8 (16-1/2°)
Edge	4° each end
Length (iron)	75"
Field	7 kG
Number of Quadrupoles	13
Aperture	8" ϕ
Length (iron)	24"
Field Gradient	1 kG/inch

Vacuum

Max. Pressure	10^{-9} Torr
Chamber Inside Dimensions	7" × 2.6" (in dipoles)

pulses would be accumulated in the ring (of circumference = 1/6 C_{AGS}) and then injected into the AGS via a synchronous bucket to bucket transfer. As only one of ten LINAC pulses/sec is currently used by the AGS, the gain in intensity would be a factor 20. This project which includes some 18 dipoles and 36 quadrupoles is expected to cost about $6M. Now of more immediate interest to this conference is the fact that for a modest additional investment this accumulator can be upgraded to a booster which would allow a ten-fold increase in the unpolarized AGS beam. This would, of course, directly benefit the type of physics I have been discussing and also offer the possibility of simplifying the CBA injection. In order to circumvent the present space charge limit in the AGS one needs to accelerate from 200 MeV to 1 GeV in the accumulator/booster. The booster would be cycled at 10 Hz with the resulting 1 GeV 10^{13}-proton bunches accumulated in the AGS. Now, of course, in order to accelerate, extract, and use such an intense beam there would need to be a number of other improvements and additions made to the AGS beyond the addition of the accumulator ring. The most important of these are included in the following list,

1. Add significant accelerating capability to the A/B RF system.
2. Modify A/B ring magnet power supplies to permit 10 Hz operation.
3. Modify the A/B extraction system magnet and power supply for 10 Hz operation.
4. Transfer line from A/B and AGS injection system must handle 1 GeV.
5. Upgrade AGS vacuum from 10^{-7} to 10^{-8} Torr, so that 1 second dwell time is possible.
6. New RF system to accelerate 10^{14} protons/pulse. Total RF (A/B + AGS) > 4 MW.
7. Radiation harden ring components, and add fast abort system.
8. Modification to beam lines and external areas.
9. Add a stretcher ring to ease extraction and raise duty cycle to 1.

The result of these efforts would be an AGS with 10 times higher average intensity than at present, with duty cycle of 1. As a side benefit the A/B would allow the construction of a 1 GeV/c π test beam which could be available year round.

Finally I'll discuss briefly the heavy ion proposal[1] which advocates yoking the AGS to the BNL Tandem Van de Graaf via a cyclotron (see Figure 12). The cyclotron would be constructed from components of the SREL cyclotron which are already on the BNL site.[20] Eventually the Tandem (.008 A GeV) will feed the cyclotron (.15 A GeV) which in turn will feed the AGS yielding up to 15 A GeV for heavy ions. Ions up to A ~ 130 (iodine) can be fully stripped

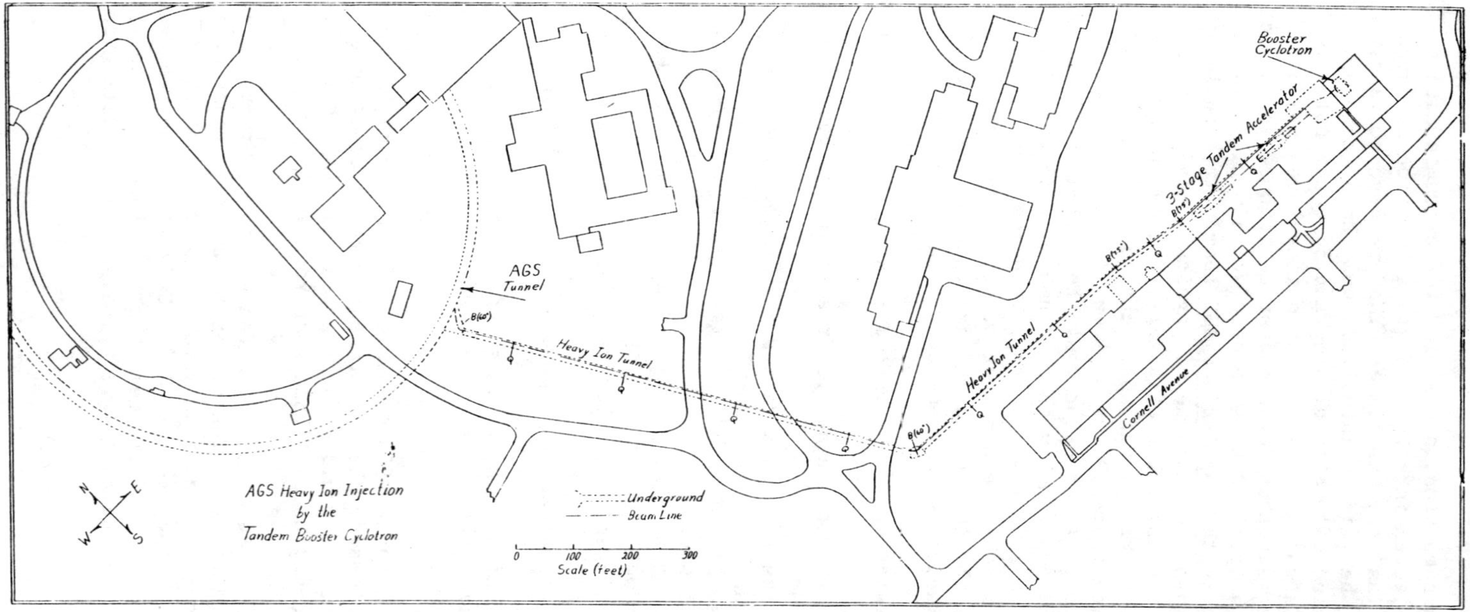

Fig. 12. AGS heavy ion injection by the Tandem booster cyclotron.

yielding currents of 10^8-10^{10} particles/pulse. Many combinations and permutations are possible. For example, without the cyclotron one can inject and accelerate ions up to S^{32} in the AGS. Table 6 shows sample AGS currents with and without the cyclotron. Without the AGS, the cyclotron can feed partially stripped ions as heavy as U^{238} back to the Tandem.

The project would take about four years and $38M. The Tandem-AGS combination could be ready in 2 years. The cost of one new and one upgraded high energy detector are included in the above estimate. A 10 week AGS program is envisioned. It is anticipated that at the energy density attainable at this facility, the hypothesized QCD phase change will occur, converting the collection of colliding nucleons into a quark-gluon plasma. If this is true a whole new field of physics will be opened up.

Ultimately the heavy ions could be injected into the CBA. Figure 13 shows the energy vs mass range attainable by various stages of the project compared to that of the Bevalac.

In summary, it looks like quite a busy future for the AGS. Even without improvements at least one generation of rare K decay experiments beyond those currently launched seems feasible. Beyond that a major effort at any of the experiments discussed above could take it to the point where it would be limited by intrinsic physics background. To pursue a full program of physics at this level one would want to increase the intensity of the AGS as described above. A ten-fold increase in K flux would remove such experiments from the category of all-out technological assaults and render them manageable by reasonably small groups of physicists. In addition, certain other, cleaner experiments, e.g., $K^0{}_L \rightarrow e^+e^-$ or $e^+e^-\pi^0$, could be pushed to limits unobtainable at the present AGS. The increased flux would also be welcomed by the neutrino and hypernuclear physics programs. Even experiments which do not at present require higher fluxes would benefit through the availability of purer beams and cleaner conditions.

The advent of polarized beams and of heavy ion acceleration will open up new whole areas of physics at the AGS. A 52 week program will be hardly enough.

Thanks for discussions and other help in preparing this talk go to J. Greenhalgh, R. Lanou, D. Lazarus, W. Louis, D. Lowenstein, M. Marx, M. Murtagh, M.P. Schmidt, L. Smith, and Y. Suzuki.

This work is supported in part by the U.S. Department of Energy under Contract No. DE-AC02-76CH00016.

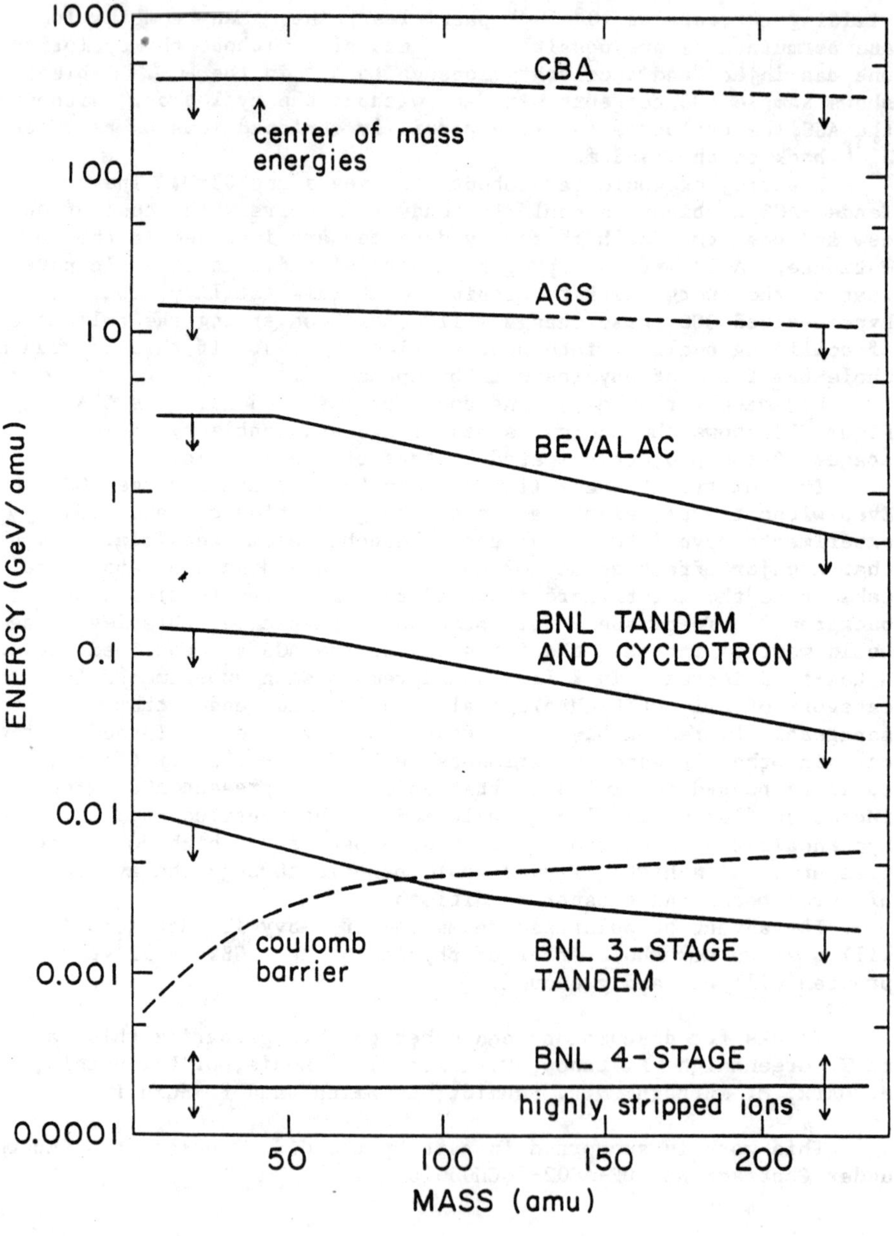

Fig. 13. Energy/nucleon vs mass for the BNL heavy ion accelerator complex and the BEVALAC. Future possible CBA performance is indicated.

TABLE 6 Expected AGS Currents

	Tandem Only		Tandem-Cyclotron	
	Oxygen	Sulfur	Sulfur	Iodine
Source	100 μa	200 μa	200 μa	200 μa
Pulsewidth	22 μsec	22 μsec	4.9 μsec	7.3 μsec
Duty Cycle	1.5×10^{-5}	1.5×10^{-4}	1.0×10^{-4}	1.6×10^{-4}
Terminal Stripper Efficiency	40%	20%	20%	20%
Tandem 2nd Stripper Efficiency	---	25%	---	---
Tandem Transmission	75%	75%	75%	75%
Buncher Efficiency	---	---	60%	60%
Cyclotron Stripper Efficiency	---	---	16%	16%
Cyclotron Acceleration and Extraction Efficiency	---	---	70%	70%
Transport to AGS	100%	100%	100%	100%
Stripping Efficiency at AGS	100%	6%	100%	30%
Number of Pulses Injected in AGS	10	10	33	33
AGS Current	3.9×10^{10} ppp	5.9×10^{8} ppp	1.8×10^{9} ppp	8.3×10^{8} ppp
AGS Energy	14.6 A•GeV	14.6 A•GeV	14.6 A•GeV	12.0 A•GeV

REFERENCES

1. "Proposal for a 15A GeV Heavy Ion Facility at Brookhaven" BNL 32250, January 1983.
2. "Proposal for an Accumulator/Booster for the AGS", March 1983, BNL 32949.
3. "High Energy Physics Achievements at the AGS and Cosmotron", R.H. Phillips, BNL 1981. "AGS 20th Anniversary Celebration", N.V. Baggett, BNL 1980.
4. K.M. Terwilliger, et al., IEEE Trans. Nuc. Sci., NS-28, No. 3, June 1981.
5. AGS Proposal 734, BNL/Brown/Pennsylvania/KEK/Osaka/Stony Brook/Tokyo (INS).
6. R.C. Larsen, et al., AGS Proposal 780, 1982.
7. D.M. Lazarus, et al., AGS Proposal 777, 1981.
8. M.P. Schmidt, private communication.
9. F. Wilczek, Phys. Rev. Lett. 49, 1549 (1982).
10. G. Bozoki, et al., AGS Proposal 776, 1981.
11. R.C. Larsen, et al., AGS Proposal 749, 1981.
12. J. Greenhalgh "On Estimating the K^0_L Flux at BNL, 1982 (unpublished).
13. CBA Newsletter No. 5, April 1983.
14. L. Littenberg "Rare K Decays", Proceedings of the Second LAMPF II Workshop, Los Alamos, NM, July 1982, p. 728.
15. M.E. Zeller, et al., BNL Summer Study on AGS Utilization 1970, p. 193.
16. D.M. Lazarus, BNL-21728, 1976.
17. M. Ferro-Luzzi, et al., CERN/PSCC 82-24 and 82-95 (1982).
18. Y. Asano, et al., Phys. Lett. 107B, 159 (1981).
19. The transient digitizers used in the experiment of Ref. 17 cost ~ \$25,000 a channel! Although it's probable that a cheaper solution can be found, one cannot multiply these entities indefinitely.
20. Alternatively the A/B, if built, could serve as the heavy ion booster, see AGS Tech. Note 187.

HYPERON AND HYPERNUCLEAR PHYSICS WITH INTENSE BEAMS

B. F. Gibson
Theoretical Division, Los Alamos National Laboratory
Los Alamos, New Mexico 87545

ABSTRACT

A brief examination of progress in the study of hypernuclear physics and the hyperon-nucleon interaction is presented. The use of Λ-hypernuclei in the study of conventional (nonstrange) nuclei is explored. The status of the hyperon-nucleon force problem is reviewed. Anecdotal results are discussed for baryon numbers 4 and 13. Σ-hypernuclei are discussed. Production of S = -2 hypernuclei is mentioned.

One of the fundamental questions facing physicists today is that concerned with how we unify the basic forces of nature: gravitational, electromagnetic, strong nuclear, and weak nuclear. Although headway has been made toward an answer, the candidate "Grand Unified Theories" are so far just that, candidate theories. Along this path, nuclear physics has contributed to our overall knowledge of the strong force; it is in a position to contribute data on the weak force. Another fundamental question facing physicists today concerns our understanding of the structure of nuclei. These multibaryon systems comprise much of the mass and energy of our immediate surroundings. Synthesis of the elements is crucially based upon nuclear structure. Nuclei produce the energy of our solar system. Their interactions involve all the forces of nature. To comprehend our universe, we must understand the structure of nuclear systems. But there exist various levels of understanding. Just as one would not attempt to study liquid argon to learn about QED, one does not expect to extract significant knowledge about QCD from studying the binding of the neutron and proton to form deuterium. Likewise, one does not attempt to calculate the structure of complex crystals starting from first principles and QED; solid state is an interesting and viable field of physics independent of quantum electrodynamics.

Particle physics seeks to provide an understanding of elementary particle interactions at very high energies (ultra short distances). In contrast, nuclear physics strives to describe the nucleus at energies and interparticle distances corresponding to conditions which one might describe by two bags barely overlapping. Here, in a region that the particle physicist finds difficult to describe quantitatively with asymptotically free theories, the nuclear physicist finds simplification and order in terms of meson exchange models. It is the possibility of speculating about the transition from the remarkably successful picture of the nucleus as a composite system of interacting nucleons to one of a quark soup that intrigues many physicists. However, one must first define the limits of validity for describing nuclear phenomena in terms of physically observable baryons and mesons before evidence for quark degrees of freedom in nuclei can be critically evaluated. Recall two successes of nuclear physics in the last decade: 1) the perfection of model calculations based solely upon nucleon degrees of freedom to the point that comparison of results with experimental data revealed the inadequacies of the assumption and demonstrated the undeniable need to expand the model to include meson exchange currents - a new degree of freedom; 2) the perfection of realistic nucleon-nucleon potential model calculations to the extent that a comparison of results with well established experimental binding energies revealed discrepancies that could only be accounted for by the introduction of three-body forces. In both cases detailed, precision calculations were required in comparison with numerous experimental data before it became possible to establish that these small but significant effects were genuine. Thus, nuclear physics seeks the appropriate degrees of freedom with which to describe nuclear systems and their interactions. The ultimate test of our intellect is whether we possess the capability to calculate all of the nuclear phenomena which we have the ability to measure.

In what follows, I will specialize the discussion to hypernuclear physics - those multibaryon systems in which one or more of the nucleons has been replaced by a hyperon (Λ, Σ, Ξ, Ω). Along the way, you will find mention of hypernuclear properties with possible relevance to quark model predictions - the Λ and Σ spin-orbit forces. You will see reference to the use of a nuclear target to search for the di-Λ (or "H" particle). These are the topics which may be most exciting to this audience. However, the primary purpose of this discourse is to impart some of the enthusiam which nuclear physicists feel for this budding subfield - to survey the interesting directions of research which would be open if there were available an intense source of kaons.

Nuclear physicists strive to understand conventional nuclear matter; they also seek to create and study new forms of quasi-nuclear matter. The K and $\bar{K}$ mesons are useful for both purposes. Our knowledge of the stucture of conventional nuclei can be enhanced by utilizing the K^+ probe.[1] Because of its strangeness (S = +1), the low energy KN interaction is not resonant. There are no known S = +1 baryons or low-lying resonances. The heavy mass of this weakly interacting hadronic probe makes it an ideal high momentum transfer tool below meson production threshold. Because it interacts with the neutron as well as the proton, the K^+ should be useful in determining the neutron's role in collective excitations and the neutron components of particle-hole states. The (K^+,K^0) charge exchange reaction should be even better suited to structure studies than the standard (π^+,π^0) reaction; the kaon distortion in initial and final states is much less than that suffered by the pion. Of even more interest is the study of hypernuclei by means of the (K^-,π) reaction. One can explore the modifications of nuclei that occur when a distinguishable baryon is inserted. Hypernuclei offer an expedient means of looking beyond that found in nature, to investigate a new form of matter containing strange quarks. The study of such strange particle matter will add a third dimension to our microscopic picture of nuclear structure.

The study of hyperon behavior in nuclear matter and the fundamental properties of hypernuclei have been, since 1953, the driving interest in $\bar{K}$-nucleus physics. That interest should soar with the advent of intense kaon beams. The $\bar{K}$-meson (strangeness S = -1) interacts very strongly with nuclei. Like the pion, the $\bar{K}$ is strongly absorbed by the nucleus; its elastic channel wave function is localized primarily in the nuclear periphery. One can understand the resonance structure in the $\bar{K}N$ amplitude in terms of the conservation of strangeness, a basic symmetry of the nuclear strong force. At threshold the open inelastic channels are: $\bar{K}N \rightarrow \pi Y$ ($Y = \Lambda$ or Σ, baryons having S = -1). The $\bar{K}$ can fuse with the nucleon to form a variety of Y* resonances (S = -1) at laboratory momenta below 1.5 GeV/c just as the π coalesces with the nucleon to form the N*'s [the $\Delta(3,3)$, etc.]. Two of the more interesting Y* resonances are the $\Lambda(1405)$ and the $\Lambda(1520)$. The $\Lambda(1405)$ lies just below threshold in the K^--atom (zero energy) system and qualitatively alters the I = 0 $\bar{K}N$ amplitudes in the nuclear medium. The $\Lambda(1520)$, with its extremely narrow width ($\cong$ 16 MeV), is potentially useful in investigating the intriguing problem of the propagation of an isobar within the nucleus. Answers to questions of how the energy and lifetime are modified due to Fermi motion, Pauli blocking, and collision damping are fundamental to our understanding the mechanism of meson propagation and the role of mesonic degrees of freedom in nuclear matter.

The (K^-,π) strangeness exchange reaction can be exploited to investigate the S = -1 Λ-hypernuclei and Σ-hypernuclei as well as the generalized Y*-hypernuclei. The (K^-,K^+) double-strangeness-exchange reaction can be used to produce the S = -2 Ξ-hypernuclei and double-Λ- or double-Σ-hypernuclei. Only the (K^-,π^+) reaction forms a unique hypernucleus (the Σ^--hypernucleus) assuming a single-step strangeness-exchange reaction mechanism. Thus, knowledge of all final state channels is required for a complete picture of the strangeness exchange reactions, in particular the isospin structure. However, nuclear structure information can be extracted from binding energies, γ de-excitation energies, angular distributions of differential cross sections, etc., even in the absence of a complete knowledge of all reaction channels.

The use of the Λ as a probe of the properties of conventional (S = 0) nuclei is one of the strongest motivating factors in our study of hypernuclei.[2] Coupling a Λ to a nucleus will change the moment of inertia of a deformed nucleus and produce a corresponding effect upon the rotational band structure; it should produce an observable effect in the phonon spectrum of a vibrational nucleus; and it may alter the energy gap in a superfluid nucleus. Near the mass regions showing oblate-to-prolate phase transitions, the addition of a hyperon may alter the mass value A at which the transition occurs. An added Λ would certainly influence the fission process and most likely the properties of shape isomers. Giant resonance properties may be altered due to coupling of a Λ to the nucleus. Core polarization induced by a Λ would alter the moments of nuclei deduced from γ transitions. Compression due to the presence of a Λ will increase the Coulomb energy of the core nucleus. Finally, the addition of a Λ to a nucleus can raise the threshold for particle emission making low-lying continuum states stable against particle decay. The insertion of a tagged baryon into the nucleus permits us to perturb the nuclear core of the hypernucleus being investigated in a way not possible by means of standard isotope or isotone studies. Each of these perturbative alterations in the nuclear core provides a different test of our understanding of the underlying nuclear structure principles.

As an example, let us consider ${}^7_\Lambda$Li, where the observation of a hypernuclear γ ray has demonstrated that the low-lying continuum levels in ^{6}Li do become particle stable.[3] The ^{6}Li nuclear core is difficult to model. There are no bound ^{5}He or ^{5}Li nuclei from which it can be formed by addition of a nucleon; it is not well represented as a hole in ^{7}Li. Thus, isotope or isotone studies do not provide realistic tests of our nuclear models of ^{6}Li. The first excited state in the ^{6}Li spectrum (see Fig. 1) lies in the continuum, above the threshold for α+d decay. Because our methods of treating continuum states differ from bound state calculational methods and are not as reliable approximations, we have been limited to comparison with ground state properties of ^{6}Li for stringent tests of our mathematical models of that nucleus. However, the

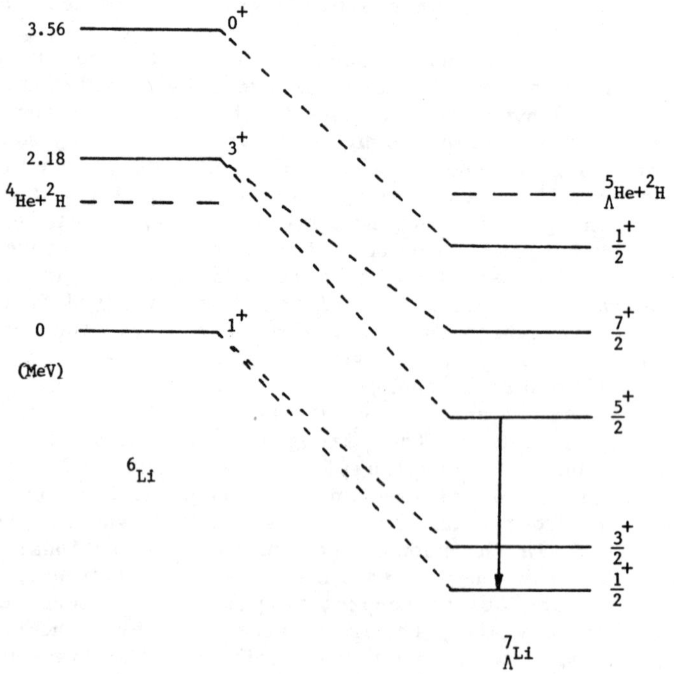

Fig. 1 Spectrum of ^{6}Li with possible corresponding particle stable levels in ${}^7_\Lambda$Li.

addition of a Λ to form $^{7}_{\Lambda}Li$ yields a hypernucleus which can be used to test our understanding of ^{6}Li. With the insertion of the relatively weakly interacting Λ, the ^{6}Li core remains intact while several continuum levels become particle stable. Our models can then be evaluated in terms of how well the dynamics of a system with several bound levels is reproduced. Our success in describing the spectrum of $^{7}_{\Lambda}Li$ depends crucially upon our correct modeling of ^{6}Li.

To understand and utilize Λ-hypernuclei, we must have a reasonable description of the ΛN interaction. The coupling of the ΛN-ΣN system in the $T = \frac{1}{2}$ channel is a complication not arising in low-energy nucleon-nucleon scattering. Experimental data on YN scattering are scarce. Because of the short lifetimes (of order 10^{-10} sec. or less) experiments are difficult, especially at low energy. Present fluxes of hyperons are not adequate to measure hyperon-nucleon cross sections. The limited low-energy YN data show a dominant s-wave character.[4] Only through the angular distributions for $\Sigma^{-}p \rightarrow \Lambda n$ inelastic scattering have nonnegligible p-wave contributions been established. At higher energies in the Λp system, near the $\Sigma^{+}n$ threshold, the data show evidence for the existence of at least one ΛN resonance (M = 2919 MeV) with a narrow (< 10 MeV) width. The lack of YN data has led to the construction of potential models very dependent upon sizeable theoretical input. More extensive data are clearly essential: to adequately treat hypernuclear structure, to verify the existence of bag model predictions of S = -1 dibaryon states, and to explore such questions as whether the short range repulsion in the nucleon-nucleon force is the result of Pauli principle effects due to the quark structure of nucleons. (If the energetically most advantageous quark configurations are forbidden, then the presence of a strange quark in the YN interaction should reduce the repulsion compared to the NN interaction.)

Data on the A=4 Λ-hypernuclear isodoublet provide a good test of the low-energy characteristics of the fundamental hyperon-nucleon force as well as a unique opportunity to study the complications that arise in calculations of the properties of systems in which one baryon (here the Λ) couples strongly to another (the Σ) with a different isospin. In particular, when one represents the free YN interaction in terms of one-channel effective ΛN potentials, the resulting 0^{+} (ground) states and 1^{+} (excited) spin-flip states of the A=4 system are inversely ordered in terms of binding energies, the 1^{+} state being more bound. However, utilizing a coupled ΛN-ΣN separable potential model, we have been able to demonstrate that the spin-isospin suppression of the Λ-Σ conversion due to the composite nature of the $^{4}_{\Lambda}H$ and $^{4}_{\Lambda}He$ systems is sufficient to yield a 0^{+}-1^{+} binding energy difference in approximate agreement with the experimental measurement, when an exact four-body formalism is used as the basis for the numerical computation.[3] That is, the $T=\frac{1}{2}$ ^{3}H and ^{3}He nuclear cores do not interact with the Λ-Σ system in the same manner as do free $T=\frac{1}{2}$ protons and neutrons; the composite nature of the trinucleon bound states suppresses the Λ-Σ conversion process in a physically observable way.

To fully develop a picture of strange particle matter, we must understand the crucial aspects of hypernuclear structure. Of particular importance are the spin and parity of levels [using the (K^-,π) angular distribution], the isospin composition of levels [comparing (K^-,π^-) and (K^-,π^0) angular distributions], the nature and strength of the residual interaction experienced by the Λ (conventional analysis of hypernuclear spectroscopy), and the effects of charge symmetry breaking in the ΛN force (comparing levels in mirror hypernuclei). To progress beyond our present rudimentary knowledge, we require much better data (to deduce, for example, a reliable parameterization of the fundamental YN force from the anlysis of hyperon-nucleon scattering data). More intense beams and better resolution than presently available are needed in order to fully utilize these tagged baryon systems, to develop our knowledge of new forms of matter as well as conventional nuclear stucture.

What are our present experimental capabilities? The known momentum transfer characteristics of the (K^-,π) reaction are shown in Fig. 2. At the "magic momentum" of about 530 MeV/c for Λ production and 280 MeV/c for Σ production, the 0° momentum transfer vanishes in hyperon production at rest within the nucleus.[5] In this momentum transfer range the production of low-spin substitutional states, in which a nucleon is replaced by a hyperon in the same orbit, is emphasized. Higher spin states emerge at nonzero angles.

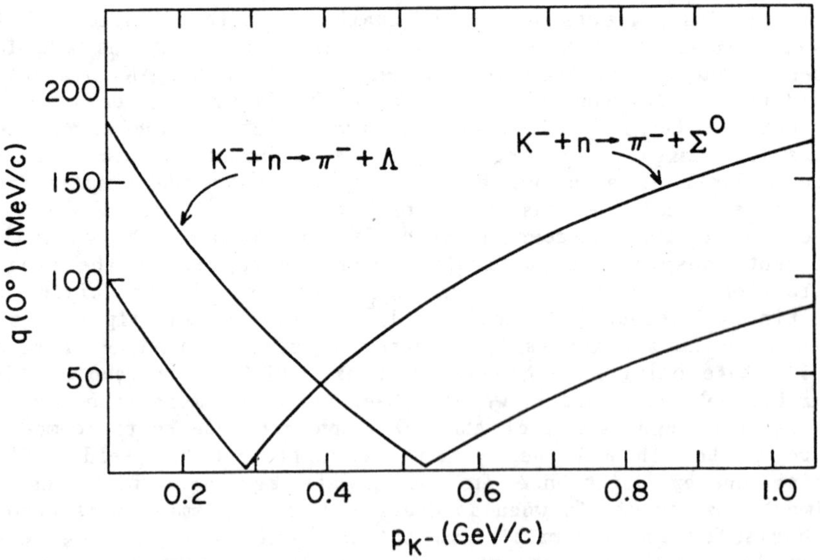

Fig. 2 Lab momentum transfer q at θ=0° as function of incident lab momentum for Λ and Σ production; large A is assumed and binding energy effects are neglected. [From C. B. Dover, L. Ludeking, and G. E. Walker, Phys. Rev. C 22, 2073 (1980).]

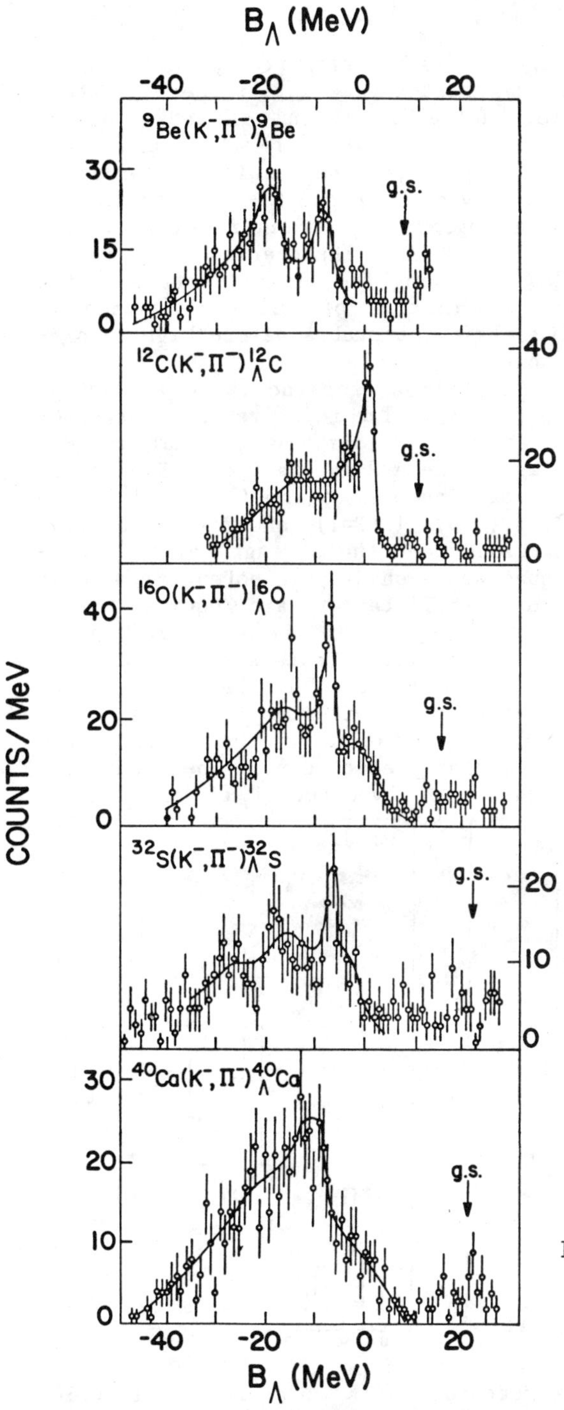

Fig. 3 Spectra for the (K^-,π^-) reaction as a function of the Λ binding energy. [From H. Brückner, et al., Phys. Lett. 62B, 481(1976).]

For example, in the (K^-,π^-) angular distribution from p-shell, spin-zero targets, the 0^+ hypernuclear states peak at 0°; the 1^- states, at about 10°, etc. As in other nuclear reactions, the shape of the angular distribution provides a clear signature for the spin of an isolated hypernuclear state. A sample from the results of the first (K^-,π^-) survey experiments is given in Fig. 3. The excitation functions are all for 0° (pion angle) and for an incident K^- momentum in the range from 700 to 800 MeV/c. The coarse energy resolution (3 to 5 MeV/c) precludes resolving the fine structure in the hypernuclear spectrum and is reminiscent of the early stage in nuclear structure physics with classical probes before high resolution spectrometers were available.[6]

More recently, angular distributions for the (K^-,π^-) reaction have been measured at BNL (see Fig. 4). The relative intensities of the peaks change with angle, and energy shifts occur that are directly related to the properties of the Λ-N interaction. Deviations from a weak coupling picture [coupling a Λ to the $0^+(T=0)$ ^{12}C core ground state plus the $2^+(T=0)$, $1^+(T=0)$, $1^+(T=1)$, and $2^+(T=1)$ excited states of ^{12}C] provide information about the strength of the spin-orbit Λ splitting and the ΛN quadrupole-quadrupole potential.[7] High resolution data on a variety of p-shell targets are required before one can sort out the details of the spin-spin and spin-orbit parts of the ΛN force. However, a very exciting feature of the data is the indication that the spin-orbit force felt by the Λ in the nucleus is very small, a surprising contrast with the large spin-orbit force felt by nucleons. The large deviation of the ratio of the sizes of the dominant peaks from that predicted using neutron pick-up strengths confirms the tendency of hypernuclei to form states

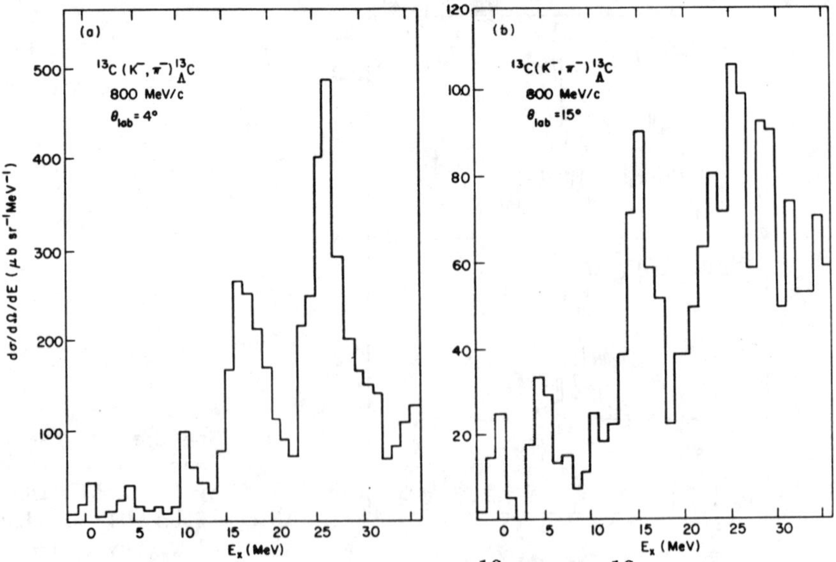

Fig. 4 Spectra for the reaction $^{13}C(K^-,\pi^-)^{13}_{\Lambda}C$ as a function of the excitation energy. [From M. May, et. al., Phys. Rev. Lett. 47, 1106 (1981).

with a higher degree of spatial symmetry than is possible in normal nuclei; if one uses as a basis the states with [54] and [441] symmetry, the [54] symmetry in ${}^{13}_{\Lambda}C$ is forbidden in a system of 13 nucleons by the Pauli principle.[7] Thus, evidence for a dynamical selection rule emerges. But the full exploitation of structure information available from the spectra of Λ-hypernuclei requires considerable improvement in energy resolution, which is possible only with more intense K^- beams.

Σ-hypernuclei studies lie at the forefront of current hypernuclear investigations.[3] Surprisingly narrow Σ states have been reported. (A large width due to strong $\Sigma \rightarrow \Lambda$ conversion had been anticipated.) Forward production of Σ's was studied in p-shell targets from Li to C using the $(K^-,\pi^\pm)$ reactions at 720 MeV/c. The best evidence was for ^{9}Be (Fig. 5); data for the production of Λ-hypernuclei are shown for comparison. Narrow Σ states have been seen since at 400 and 450 MeV/c, presumably corresponding to coherent substitutional transitions leading to 0^+ final states. Interesting questions arise in the interpretation of these data. Why are some Σ states relatively narrow? What are the single-particle properties of a Σ in the nucleus (well depths, spin-orbit potentials, etc.)? Do Σ states have good isospin? The data are yet too crude to permit definitive answers. However, there are tantalizing hints that the Σ spin-orbit potential is larger than that of the

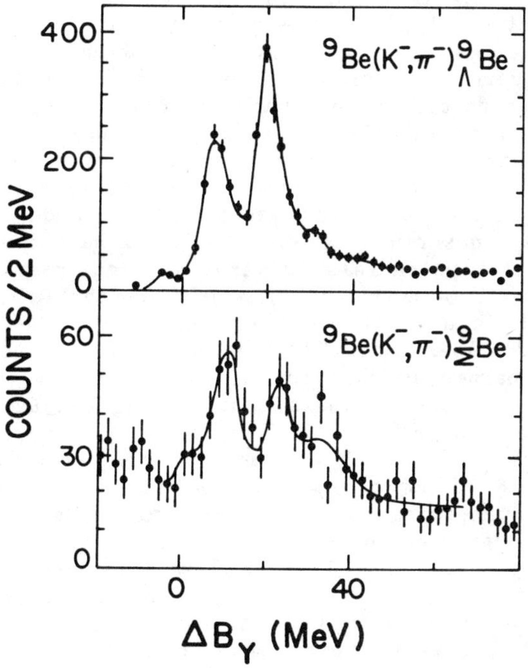

Fig. 5. Spectra for the (K^-,π^-) reaction on ^{9}Be leading to Λ-hypernuclei and Σ-hypernuclei. [From R. Bertini, et al., Phys. Lett. 90B, 375 (1980).]

nucleon; recall that for the Λ it appears to be almost zero. Angular distributions for both the (K^-,π^+) and (K^-,π^-) reactions are needed to answer these questions, as well as to obtain definite spin assignments. An intense, low momentum K beam would be of immense benefit in the study of Σ-hypernuclei.

Hypernuclear physics utilizing the double strangeness exchange reaction (K^-,K^+) lies in the future.[7] Cross sections for nuclear targets will be quite small (a few nb/sr to a μb/sr). Thus, the study of S = -2 hypernuclei would benefit enormously from the availability of an intense kaon beam in the 1- to 2- GeV/c momentum range. The spectroscopy of Ξ- and $\Lambda\Lambda$-hypernuclei represents a logical progression in the evolution of hypernuclear physics. The spectroscopy of such hypernuclei is rich, although only a restricted set of states (high spin with no spin-flip transitions) will be excited with measurable cross section using the high momentum transfer (K^-,K^+) reaction. Determining single particle properties of the Ξ is one goal; exploring the $\Lambda\Lambda$ interaction is another. The $\Xi^- p \rightarrow \Lambda\Lambda$ transition is not expected to broaden the levels significantly beyond what has been seen in Λ- and Σ-hypernuclei. The $\Lambda\Lambda$ pairing correlation effects should enhance states in $\Lambda\Lambda$-hypernuclei just as NN correlations do in S = 0 nuclei.

As remarked previously, the search for evidence of quark degrees of freedom in nuclear matter is a quest of current interest to many physicists. Let me remind you that we have already found them in the sense that one believes the quark description of N's and Δ's; i.e. $NN \rightarrow N\Delta$ excites a new quark degree of freedom. Likewise $\Sigma N \rightarrow \Lambda N$ involves a quark transition. However, there are two areas where hypernuclear physics offers some hope of providing credible evidence of a positive nature for the bag approach to quark models. First, the disparate sizes of the experimentally oberved mean-field Λ-nucleus spin-orbit force and Σ-nucleus spin-orbit force may differentiate between quark-model and meson-exchange model descriptions. Naive quark model descriptions of the Λ and Σ have led to ΛN and ΣN two-body spin-orbit potentials of very different magnitudes.[3] However, the step from two-body spin-orbit potential to one-body, mean-field force in a shell model is not a short one. Second, the search for the doubly strange "H" dibaryon, first proposed by Jaffe, is clearly of paramount importance. The $^3He(K^-,K^+n)$"H" reaction would appear to be the cleanest test of the existence of such a massive six quark object as is predicted by some quark models. Particle physics seeks at high energies the asymptotic, small r limit of particle phenomena, in contrast to nuclear physics where one goes to low energies to find asymptopia. As nuclear physics moves up in energy and momentum transfer to find new degrees of freedom and as particle physics moves down in energy to seek structure information beyond the r=0 limit, there is hope that these two once common fields will again come together.

The work of B.F.G. was performed under the auspices of the U. S. Dept. of Energy. He gratefully acknowledges informative conversations with J. D. Walecka, C. B. Dover, E. M. Henley, J. L. Friar, and G. E. Brown.

References

1. Proceedings of the Workshop on Program Options in Intermediate Energy Physics, compiled by J. C. Allred and B. Talley, LA-8335, Vol. 1, pp. 139-143 (August 1979).
2. H. Feshbach, Meson-Nuclear Physics 1976, editied by P. D. Barnes, R. A. Eisenstein, and L. S. Kisslinger, AIP Conf. Proc. No. 33, p. 521.
3. Proceedings of the International Conference on Hypernuclear and Kaon Physics, MPIH-1982, Vol. 20, ed. by B. Povh.
4. J. J. de Swart, Nukleonika 25, 397 (1980).
5. B. Povh, Annu. Rev. Nucl. Part. Sci. 28, 1 (1978).
6. Proceedings of the 1979 International Conference on Hypernuclear and Low Energy Kaon Physics, Nukleonika 25, nos. 3, 4, and 9 (1980).
7. C. B. Dover and G. E. Walker, "The Interaction of Kaons with Nucleons and Nuclei," Phys. Rep. C 89, 1 (1982).

ISSUES IN THE STANDARD MODEL

Mary K. Gaillard† ††
Fermi National Laboratory, P.O. Box 500, Batavia, IL 60510

ABSTRACT

Focussing on the standard electroweak model, we examine physics issues which may be addressed with the help of intense beams of strange particles.

INTRODUCTION

I was assigned the topic "issues in the standard model," in so far as they are relevant to high intensity sources of strangeness. It is not really clear what is meant by the "standard model" in this context, and, obviously, one of the most important issues in the "standard model" is testing it--in other words, looking for non-standard effects. So I have collected a miscellany of issues, starting with some philosophical remarks on how things stand and where we should go from here. I will then focus on a case study: the decay $K^+\to\pi^+$+nothing observable, which provides a nice illustration of the type of physics that can be probed through rare decays. Other topics I will mention are CP violation in K-decays, hyperon and anti-hyperon physics, and a few random comments on other relevant phenomena.

PHILOSOPHY

One might claim that things have never been better in high energy physics. We have finally achieved a longstanding goal: The elaboration and successful testing of a renormalizable theory of the weak, electromagnetic and strong interactions. We even have indications, specifically the value of the neutral current parameter $\sin^2\theta_W$, that these interactions are unified in a "grand" renormalizable theory. (The response to all this success is, of course, that renormalizability--the erstwhile holy grail--is no longer "in," and many theorists are now working on non-renormalizable theories!)

† On leave of absence from Department of Physics, University of California, Berkeley and Lawrence Berkeley Laboratory.

†† Supported in part by the National Science Foundation under Research Grant No. PHY-82-03424.

One may also argue that things have never been worse. No one believes that the above theories provide the ultimate description of nature. We want to solve the gauge hierarchy problem, understand fermion masses, superunify, find quark and lepton substructures...the problem is that we have gotten ahead of ourselves. There are no data to guide us along these roads--not even monopoles, and as yet few decaying protons. This leaves the way open for wild speculation, which is fun, but doesn't necessarily represent progress.

We clearly need to probe energies higher than those presently accessible in the laboratory. The standard attack in this direction is three-fold:

1) Cosmology. The Big Bang provides the highest energy laboratory around, but the data are not always easy to interpret since they came from a single event in experimental conditions not controlled by us.

2) Let $E_{Lab} \to \infty$. In real life, of course, infinity will be replaced by some practical cut-off Λ_{pr} which is possibly 10's of TeV, but not many orders of magnitude more.

3) Precision measurements at "low" energies: $E_{Lab} \ll \Lambda_{pr}$. The prime example of this approach is the proton decay search which we believe probes energies up to 10^{14} or 10^{15} GeV. As an example more relevant to this workshop, suppose there were a direct "generation changing" interaction mediated by a boson of mass m_x and coupling with the usual semi-weak strength. Depending on the branching ratios accessible, rare decay searches might probe beyond 10's of TeV, as can be seen by parameterizing some typical branching ratios in terms of m_x:

$$B(K_L \to \mu e) \sim 10^{-12}(100\ \text{TeV}/m_x)^4,$$

$$B(K_L \to \pi^0 \mu e) \sim 10^{-12}(170\ \text{TeV}/m_x)^4, \tag{1}$$

$$B(\Sigma^+ \to p \mu e) \sim 10^{-7}(\text{TeV}/m_x)^4 .$$

There are, in addition, still things to be learned about physics at more modest energies. For example, we still don't know how to calculate low energy hadronic matrix elements. Perhaps the confinement/lattice theorists will resolve this difficulty, but new experimental input could certainly be of help. Can high precision measurements and studies of rare processes instruct us on this issue? We are still in the dark concerning the origin of CP violation. Will experiments eventually reveal some small deviation from the superweak predictions?

In discussing these questions in more detail, I will adopt for the most part a desert scenario. The justification for taking this desolate view point is that it provides a well defined yardstick for gauging the experimental accuracy we should aim for. The point is that even the desert has some oases. As we let $E_{Lab} \to \Lambda_{pr}$, we have still (maybe?) to uncover the top quark, for example, and we still have no experimental clue as to the nature of spontaneous gauge symmetry breaking. There is a sort of "unitarity limit"[1] of about a TeV associated with the standard electroweak theory:[2] we must find some evidence for scalar structure at an effective center of mass energy of a TeV or less. The advice I would give to high energy planners is: aim for the hardest thing to find, namely the detection of a "minimal model" Higgs boson in a mass range up to the TeV level. Then you are bound to find something, and hopefully your data will reveal a much richer structure.

By the same token, in thinking about high precision measurements: aim for those tiny effects predicted in the minimal model. If you can measure them, you will in any case learn something, and you may indeed uncover more interesting unexpected phenomena.

$K^+ \to \pi^+ +$ nothing

Following the above line of reasoning, the special interest of this decay mode is that we (almost) know it's there. The minimal model with three generations of fermions predicts a branching ratio

$$B(K^+ \to \pi^+ \nu\bar{\nu}) = 0.7 \times 10^{-10} |1 - x_t|^2, \tag{2}$$

where x_t is the top quark contribution: $x_t = 0$ corresponds to the estimate[3] for the GIM 4-quark model,[4] and

$$x_t \simeq \theta_t \left(\frac{m_t}{m_c}\right)^2 \frac{\ln(m_W^2/m_t^2)}{\ln(m_W^2/m_c^2)} \tag{3}$$

in the K-M 6-quark model.[5] Here and elsewhere we use the formulae valid for $m_t \ll m_W$, which may not be a very good approximation, but it simplifies the discussion and does not significantly affect the order of magnitude estimates we are after.[6] In Eq. (3) we have introduced the parameter

$$\theta_t = (\theta_{dt}\theta^*_{st})/(\theta_{du}\theta^*_{su}), \tag{4}$$

where the θ_{ij} are the relevant elements of the K-M mixing matrix. Since $m_t > 20$ GeV (we take everywhere $m_c = 1.5$ GeV),

$$\ln(m_t^2/m_W^2)/\ln(m_c^2/m_W^2) \lesssim 1/4 \tag{5}$$

and from the observed rate for $K_L \to \mu\mu$ Shrock and Voloshin[7] derived an upper bound which can be expressed as:

$$|\theta_t \; m_t^2/m_c^2| \leq 25. \tag{6}$$

Recent refinements[8,9] give a slightly smaller value, but I prefer to be conservative here since all estimates are rough. The main uncertainty is in the real part of the intermediate 2-γ contribution to $K_L \to \mu\mu$, although if supersymmetry is valid at relatively low energies (hundreds of GeV), there are apparently cancellations[10] which can invalidate[11] the bound (6) altogether. I shall ignore this possibility in the subsequent discussion. By using all available data including the K_L-K_S mass difference,[12] it is possible to bound[9] $|1-x_t|$ from below. However I prefer not to use Δm_K as a constraint, since there are well known uncertainties associated with this analysis. I don't think that one can exclude with certainty at present the possibility that $x_t \simeq 1$, but I consider this perversion of nature as rather unlikely. It would require rather smaller mixing angles than we expect:

$$\theta_t < (0.15)^2 \text{ for } m_t > 20 \text{ GeV, or}$$

$$\theta_t = (0.085)^2 \text{ for } m_t \simeq 35 \text{ GeV.}$$

Since θ_t is related by the unitarity of the K-M matrix to the parameters governing b-decay and ν-induced c- and b- production, precise measurements of B lifetime and branching ratios, and ν-induced multi-lepton events should be able to yield[12,13] reliable lower limits on $|\theta_t|$ and $|1-x_t|$ which are independent of the uncertainties inherent in the analysis of Δm_K.

What do we learn from a measurement of the decay $K^+ \to \pi^+ +$ nothing observable? If we know m_t and θ_t, this decay provides the cleanest available test of weak radiative corrections within the context of the standard model. Alternatively, accepting the theory as correct, a measurement of the decay rate provides an independent constraint on the parameters m_t and θ_t. Finally, if we believe the standard model calculation and have sufficient outside constraints on m_t, θ_t, the decay can be used to probe for new physics. We list some examples.

- Flavor changing currents: A direct decay mediated by a heavy boson of mass m_x would have branching ratio

$$B(K^+ \to \pi^+ \nu_e \bar{\nu}_\mu) \simeq 10^{-10} \left(\frac{26 \text{ TeV}}{m_x} \right)^4, \tag{7}$$

allowing perhaps a probe of masses up to about 25 TeV.

- Neutrino counting:[9] If there are N_ν light ($m_\nu \ll m_K$) neutrinos with the usual weak couplings (and associated leptons of mass $m_L \lesssim m_W$;

for $m_L >> m_W$ only Z^0-exchange contributes and the formula is modified, but the order of magnitude is similar) then the branching ratio for $K^+ \to \pi^+ + \nu\bar{\nu}$ is, from Eq. (2)

$$\sum_{\nu\text{-types}} B(K^+ \to \pi^+ + \nu\bar{\nu}) \simeq \frac{N_\nu}{3} (0.7) |1-x_t|^2 \leq \left(\frac{N_\nu}{3}\right) 5 \times 10^{-9} \qquad (8)$$

where we have used the bounds of Eqs. (5) and (6) to get

$$|x_t| \leq 6, \text{ or } |1-x_t| \leq 7. \qquad (9)$$

Thus a measured branching ratio exceeding a few $\times 10^{-9}$ could be interpreted[9] as signalling more than three generations of fermions. Alternatively, once we know sufficiently well the parameters θ_t and m_t so as to bound the decay rate per neutrino from below, a measurement of $K^+ \to \pi^+ + \nu\bar{\nu}$ will provide an upper bound on the number of neutrinos, within the context of the standard model.

- Neutrino masses: The branching ratio for the cascade decay $K^+ \to \pi^+ \pi^0$, $\pi^0 \to \nu\bar{\nu}$ is given by[14]

$$B\,(K^+ \to \pi^+ \underset{\hookrightarrow \nu\bar{\nu}}{\pi^0}) = 3.5\times 10^{-13} \left(\frac{m_\nu}{\text{MeV}}\right)^2 \left[1-\left(\frac{2m_\nu}{m_\pi}\right)^2\right]^{1/2} . \qquad (10)$$

This branching ratio exceeds 10^{-10} if there is a neutrino (e.g. ν_τ) with the usual neutral current couplings in the mass range $m_\nu \simeq (20-65)$ MeV. This decay is signed by a monochromatic π^+.

- $K^+ \to \pi^+$ +funnies, where the "funnies" are exotic, neutral, non-interacting particles. These could be, for example, a single spin-0 particle or a pair of fermions. This type of decay mode is really the province of the discussion by Wilczek.[15] I shall comment only on non-neutrino fermion pairs which are expected in supersymmetric (susy) theories. Some models entail a photino $\tilde{\gamma}$--the fermionic susy partner of the photon--which is very light. Calculations[14] show that for squark (scalar partners of quarks) masses above 20 GeV [the present lower limit on slepton=(scalar partner of lepton) masses is (16-19) GeV] the branching ratio is

$$B(K^+ \to \pi^+ \tilde{\gamma}\tilde{\gamma}) \leq 10^{-10} \qquad (11)$$

except for two special cases. The first of these exceptions is a photino mass range $m_{\tilde{\gamma}} \simeq (2-65)$ MeV, for which with not-too-heavy squarks one gets, via the cascade decay

$$K^+ \to \pi^+ + \underset{\hookrightarrow \tilde{\gamma}\tilde{\gamma}}{\pi^0} , \qquad (12)$$

a rate competitive with $K^+ \to \pi^+ \nu\bar{\nu}$. As for the cascade decay to neutrinos, (12) is signed by a monochromatic π^+. A second exception is the (currently most popular) class of models in which squark masses arise through radiative corrections. If we use a phenomenological Lagrangian, with a "soft" tree-level susy breaking squark mass matrix, the amplitude for $K^+ \to \pi^+ \gamma\gamma$ is logarithmically divergent,[16] the cut-off being provided by the susy breaking mass scale. In such a scenario the decay rate can be quite large, depending on how far one is willing to push up that scale.

I do not believe that photinos will be sufficiently light to be decayed into by K's in such a scenario. In fact I do not personally believe that--even if supersymmetry is relevant to physics--photinos are sufficiently light for this decay to occur in any scenario. What I do believe in is the importance--short of detecting susy partners of ordinary particles--of excluding their existence in whatever mass range is available to experiment. While there is no evidence as yet that supersymmetry is relevant to nature, there is very little evidence against it--an example of the the free rein for theoretical speculation which I alluded to above.

To summarize this case study, there are three possibilities. 1) The branching ratio for $K^+ \to \pi^+ +$ "nothing" is found to lie in the range 10^{-10}-10^{-9}. A precise measurement allows a test of radiative corrections within the minimal model and/or a measurement of the K-M mass matrix parameters. 2) The branching ratio far exceeds 10^{-9}. This signals new physics. It's interpretation lies with the discretion of the reader, but in any case the result is exciting. 3) The decay remains undetected at a level below 10^{-10} in branching ratio. This presumably also implies new physics but will be no more helpful than, for example, the lack of detection of proton decay. I optimistically consider this last possibility as unrealistic.

CP-VIOLATION IN K-DECAYS

There is no question that precision measurements in the decays $K_{L,S} \to 2\pi$ are highly desirable. They will 1) further constrain deviations from CPT invariance, a fundamental symmetry of the local Lagrangian theories which we take for granted, and 2) hopefully reveal a small deviation from the predictions of the "superweak" model, in which all CP-violating effects are in neutral meson mass mixing parameters. In discussing these effects I shall follow the dictum outlined above: ask what are the tiny effects expected in the "minimal model." If we can detect these, we can also detect grosser deviations from them, and so we are sure to learn something. Furthermore, I shall argue that if we analyze the data within the context of the minimal model, insofar as CP-violating effects are observable at all, their measurement can shed light on the still ill-understood dynamics of weak interactions, and, in particular, on the persisting mystery of the $|\Delta I|=1/2$ rule.

In the present context we understand as the "minimal model" the so-called K-M model of CP-violation in which CP-violating phases appear originally in Yukawa couplings of fermions to the Higgs particle, and, upon diagonalization of the fermion mass matrix, are shifted to the charged current (K-M) coupling matrix. As is well known, in this model observable CP-violating effects require the existence of at least three generations of fermions. As a result, any observable CP-violating effect must know about the presence of b,t quarks. To lowest order in the weak interactions, such effects occur only through "penguin" diagrams. Here we designate as "penguin diagrams" the generic class of diagrams in which the transition $d\to s$ occurs along a quark line in a bound quark (hadronic) system via $W^{\pm}$ exchange with the intermediate (u,c,t) quark system interacting with other bound quarks through gluon exchange. Within this picture we can roughly parameterize the "direct" (as opposed to superweak=mass mixing) CP-violating contribution to a decay amplitude by:

$$\frac{\mathrm{Im}A}{\mathrm{Re}A} \sim f_P x_\delta \equiv f_P\, s_2 s_3 \sin\delta \ln\left(\frac{m_t^2}{m_c^2}\right), \tag{13}$$

where f_P represents the fractional contribution of penguin diagrams to the process considered, and θ_i, $s_i=\sin\theta_i$, and δ are parameters in the K-M matrix. The combination of parameters in Eq. (13) (where we have assumed the validity of a small angle approximation) can be expressed, for example as:

$$s_2 s_3 \sin\delta \simeq -\,\mathrm{Im}\theta_{sc} \simeq \mathrm{Im}\theta_t^* . \tag{14}$$

Because penguin diagrams involve an $s\to d$ transition with I-spin conserving gluon emission, the CP-violating phase arises only in $|\Delta I|=1/2$ transitions. This leads[17,18] to a phase difference between, for example, the amplitudes for I=0 and I=2 final states in $K^0\to 2\pi$.

The superweak contribution to CP violation in neutral kaon decay arises from a $K^0\to\bar{K}^0$ term in the neutral kaon mass matrix. The CP violating parameter can be expressed as[17]

$$\varepsilon_m = \frac{\mathrm{Im Ampl.}(K^0\to\bar{K}^0)}{\Delta m_K} \simeq 2 s_2 s_3 \sin\delta \left[\ln\left(\frac{m_t^2}{m_c^2}\right) - 1 - \theta_t \frac{m_t^2}{m_c^2}\right], \tag{15}$$

using the same approximations[6] as before. For the denominator we use the original estimate[3] of Δm_K in a 4-quark flavor model, simply because this gives the right answer to within 30% for $m_c=1.5$. For the numerator, the free quark model estimate is a reasonable approximation except for the uncertainty[19] in evaluating the matrix element between kaon states of the effective quark operators. This gives an uncertainty in an overall multiplicative factor of order unity. Further strong interaction corrections[20] modify by factors O(1) the coefficients of the various terms in brackets in Eq. (15). Finally, the quantity relevant to experiment is not ε_m but

$$e^{-i\pi/4}\ \varepsilon \simeq \frac{1}{\sqrt{2}}\,\varepsilon_m + \sqrt{2}\,\xi_{2\pi,I=0} \simeq 2\times 10^{-3} \tag{16}$$

where $\xi_{2\pi,I=0}$ is the CP-violating phase in the decay of K^0 into an I=0 dipion state. In the commonly used Wu-Yang convention this phase is set equal to zero and ε_m is redefined by the shift (16). This gives an additional (small[20]) change in the coefficient of the log term in (15). For the sake of order of magnitude arguments I shall use (15) as is without corrections. The point I wish to make is simply that since

$$\ln\left(\frac{m_t^2}{m_c^2}\right) > 5 \text{ for } m_t > 20 \text{ GeV}, \tag{17}$$

the bound (6) implies that the log term in (15) contributes at least a fifth of the total magnitude. Thus we expect

$$|x_\delta| \simeq (0.2\text{-}1)\,\frac{1}{\sqrt{2}}\,\varepsilon \simeq (0.3\text{-}1.4)\times 10^{-3}\ . \tag{18}$$

In other words we expect deviations from superweak theory to occur at a level

$$|f_p x_\delta| \sim 10^{-4\pm 1} \tag{19}$$

which is the level of detection experimenters should aim for.

Two alternative optimal scenarios would be: a) Direct CP-violating effects are found at a level considerably larger than 10^{-3}. This would suggest that the standard K-M model is incorrect and signal new physics, fun and excitement. b) Effects at the expected level of 10^{-4} are measured. When m_t and the K-M angles s_1, s_2 are determined independently, the parameters f_p and δ can be extracted from the analysis of CP-violating phenomena. This will have the bonus of determining the importance of penguins and perhaps contribute to our understanding of non-leptonic decay dynamics.

But alas, as we see below, measurable effects which are proportional to $f_p x_\delta$ tend to be suppressed by other factors.

The most promising place to look for a deviation from superweak theory is still in the $K_{L,S}\to 2\pi$ decay. In this processes the deviation is characterized by the parameter ε':

$$|\varepsilon'| = \frac{1}{\sqrt{2}}\,\left|f_p x_\delta\,\frac{A(I=2)}{A(I=0)}\right| \simeq \frac{1}{25}\,|f_p x_\delta| \tag{20}$$

where the last factor includes the measured suppression of the I=2 final state relative to I=0. The present experimental limit is usually quoted as

$$\left| \frac{\varepsilon'}{\varepsilon} \right| \lesssim \frac{1}{50} , \tag{21}$$

while the above analysis suggests

$$\left| \frac{\varepsilon'}{\varepsilon} \right| \simeq \frac{1}{25} f_P \frac{x_\delta}{\varepsilon} \simeq \frac{f_P}{50} \left(\frac{1}{2} \to 2 \right), \tag{22}$$

so we expect the next round of experiments to show a non-zero effect, thus providing information on f_P.

For $K \to 3\pi$, the amplitudes are completely determined in terms of the (real, by convention) amplitude for $K \to 2\pi(I=0)$, using the $\Delta I = 1/2$ rule and chiral symmetry. Thus "direct" CP-violation can arise only to the extent that one of these is inexact, and we expect effects no larger than $10^{-1} f_P x_\delta < 10^{-4}$.

Rare K-decays which can proceed only via higher order processes with internal quark loops can have a relatively enhanced CP-violation. Unfortunately the decay rates for the interesting cases are exceedingly small. For example, the measured branching ratio for $K^+ \to \pi^+ e^+ e^-$ agrees fairly well with the (somewhat questionable in this case) estimate using free quarks.[3] The same model gives[3,17]

$$\Gamma(K_1 \to \pi^0 e^+ e^-) \simeq (K^+ \to \pi^+ e^+ e^-),$$

$$\varepsilon'_{\pi ee} = \frac{\Gamma(K_2 \to \pi^0 e^+ e^-)}{\Gamma(K_1 \to \pi^0 e^+ e^-)} \simeq x_\delta \simeq (0.2 \to 1)\ \varepsilon/\sqrt{2}, \tag{23}$$

i.e. a fairly large ε'/ε ratio, but the expected K_L branching ratio from the direct decay is only:

$$B(K_2 \to \pi^0 e^+ e^-) \sim (1 \to 5) \times 10^{-12} . \tag{24}$$

Similarly, the (here more reliable) quark model estimate gives[3,17]

$$\Gamma(K_1\to\pi^0\nu\bar{\nu}) \simeq \Gamma(K^+\to\pi^+\nu\bar{\nu})$$

$$\varepsilon'_{\pi^0\nu\bar{\nu}} = \frac{\Gamma(K_2\to\pi^0\nu\bar{\nu})}{\Gamma(K_1\to\pi^0\nu\bar{\nu})} \simeq \varepsilon\,\frac{m_t^2}{m_c^2}\,\frac{\ln(m_W^2/m_t^2)}{\ln(m_W^2/m_c^2)}\,\frac{1}{[\ln(m_t^2/m_c^2)-1-\theta_t]}$$

$$\frac{\varepsilon'_{\pi^0\nu\bar{\nu}}}{\varepsilon} \simeq (1\to10) \qquad (25)$$

where the optimistic factor 10 assumes $x_\delta \simeq \varepsilon/\sqrt{2}$, $m_t \simeq 35$ GeV. Even in this case, the $K_L\to\pi\nu\bar{\nu}$ branching ratio is not expected to exceed 10^{-13}. Of course K_L, K_S interference effects will be very pronounced in these decays. All one needs is to make a beam of 10^{13} K_S per pulse!

I list these numbers to show where minimal expectations lie. I leave it as a challenge to experimenters to attempt to measure such tiny effects, and to theorists to think of something better.

HYPERON DECAY

I submit that not much can be learned by improving experimental precision on non-leptonic decay amplitudes (aside from phases). I suspect that the present experimental errors are smaller than any conceivable accuracy theorists will ever achieve in calculating these amplitudes.

There is however some interest in improving accuracy on non-leptonic decay parameters. One would like to study SU(3) breaking corrections to the Cabbibo model and improve limits on deviations from it such as the presence of right-handed currents.[21] There is in fact a reported discrepancy[22] in $\Sigma^-\to ne^-\bar{\nu}_e$ which is yet to be resolved.

Studies of radiative decays might contribute to our understanding of non-leptonic decay dynamics. Experiments show a large SU(3)-forbidden asymmetry (with large errors) in the decay $\Sigma^+\to p\gamma$, and improved precision is needed to clarify this issue. A concerted study of the various radiative decay modes, including $\Xi^-\to\Sigma^-\gamma$, $\Xi^0\to\Lambda\gamma$, $\Sigma^0\gamma$, $\Lambda\to n\gamma$, would address the issues of "long distance" decay dynamics (penguins and all that) because the short distance contribution, i.e. the magnetic transition $s\to d+\gamma$, is highly suppressed by helicity conservation of gauge couplings. A word of warning however; the baryon pole contribution, which measures directly the weak $B\to B'$ transition, is not expected[23] to dominate over direct emission contributions in charged hyperon decay, and the limited data available[24] suggests that the same is true for neutral hyperon radiative decay. So interpretation of the data may be less than straightforward.

Finally, one can look for time reversal violation[25] by measuring the relative phase between s- and p-wave amplitudes. A deviation from the phase difference arising from strong rescattering in the final state is a sign of T-violation. Again one would want to aim for an accuracy of better than 10^{-3} in the measured phase.

ANTI-HYPERON DECAY

Comparison between hyperon and anti-hyperon lifetimes provide a test of CPT but this is unlikely to be competitive with tests provided by $\tau_{\pi}\pm$ and especially by precision measurements in the neutral kaon system.

While CPT invariance requires equal total decay rates for hyperon and anti-hyperon, CP - violation can induce differences in partial rates if there is more than one open channel and if these communicate via strong interactions. Thus for $Y\to N\pi$ there are two final state channels I=1/2, 3/2, which are eigenstates of the strong S-matrix, while the specific charge modes (e.g. $n\pi^0$, $p\pi^-$) are not. Then one gets a decay asymmetry:

$$A = \frac{\Gamma(Y\to N\pi)-\Gamma(\bar{Y}\to\bar{N}\bar{\pi})}{2\Gamma(Y\to N\pi)} \sim \sin\phi \sin\delta \frac{2|A_{3/2}||A_{1/2}|}{|A_{3/2}|^2+|A_{1/2}|^2} \qquad (26)$$

where $\delta=\delta_{3/2}-\delta_{1/2}$ is the difference between strong interaction phase shifts in the I = 3/2, 1/2 final states and $\phi=\phi_{1/2}-\phi_{3/2}$ is the difference in CP - violating phases. In the standard K-M model we expect

$$\phi_{3/2} = 0, \ |\phi_{1/2}| \simeq |f_P^{(B)} x_\delta| \qquad (27)$$

where $f_p^{(B)}$ is the fractional importance of penguins in baryon decays (generally unequal to $f_P^{(K)}$, but presumably similar in order of magnitude). Note that in addition to non-vanishing ϕ and δ, an appreciable effect depends on $A_{1/2,3/2}$ having similar strength. Herein starts the difficulty.

For[25] $\Lambda\to N\pi$, the $\Delta I=1/2$ rule suppresses the I=3/2 final state; from experiment

$$|A_{3/2}/A_{1/2}|_\Lambda \simeq 0.03. \qquad (28)$$

For the decays $\Sigma\to N\pi$ both I=3/2 and I=1/2 final states are allowed by the $\Delta I=1/2$ rule. However for p-waves the near vanishing of the $\Sigma^-\to n\pi^-$ amplitude tells us that

$$|A_{3/2}/A_{1/2}|_{\Sigma(\text{p-wave})} \simeq 0.05. \qquad (29)$$

In addition we expect $\delta << 1$ for p-waves. For s-waves the strong phase shift δ could be appreciable, and we know that $|A_{3/2}| \approx |A_{1/2}|$. However, if one believes that s-wave baryon decays are correctly described by soft pion theorems, then the amplitude for $\Sigma^+ \to n\pi^+$, which is a specific linear combination of I=1/2 and 3/2, vanishes separately for penguin and the $\Delta I=1/2$ part of non-penguin contributions. This means that

$$f_P^{(3/2)} = f_P^{(1/2)} \tag{30}$$

up to violations of $\Delta I=1/2$ and/or the soft pion limit. This in turn implies that $A_{1/2}$ and $A_{3/2}$ have equal phases to the same approximation:

$$(\phi_{3/2} - \phi_{1/2})_{\Sigma(\text{s-wave})} = O(10^{-1} f_P^{(\Sigma)} x_\delta) < 10^{-4} \tag{31}$$

Again, however, if effects as small as (31) could be detected, their measurement could contribute to our understanding of the decay dynamics. I close this section with the same challenge to theorists and experimentalists as above.

RANDOMONIA AND CONCLUSION

There is a strong theoretical prejudice that the Higgs scalar of the minimal electroweak model must have a mass

$$m_H \geq 10 \text{ GeV}. \tag{32}$$

While well founded and highly plausible, the bound (32) is not a rigorous theorem. To my knowledge the experimental bound is still

$$m_H \geq 15 \text{ MeV}. \tag{33}$$

A search[26] for

$$K^+ \to H + \pi^+, \quad H \to e^+ e^-$$

is on the edge of ruling out $m_H < 2\, m_\mu$, but the branching ratio $\lesssim 4 \times 10^{-8}$ is not quite conclusive.[27] Studies of

$$K_L \to \begin{cases} \pi^0 e^+ e^- \\ \pi^0 \mu^+ \mu^- \\ \pi^0 \gamma\gamma \end{cases}$$

at a branching ratio level of 10^{-10}-10^{-11} would eliminate with certainty the possibility that $m_H < 2m_\pi$.

The decay $K_L \to \mu\mu\gamma$ should be competitive with $K_L \to \mu\mu$ as it is lower order in α although disfavored by phase space. It could provide a laboratory for studying the $\mu^+\mu^-$ bound state.[28]

Finally, it has been suggested[29] that we should not limit our considerations to weak interactions, and that, for example, a precision measurement of fixed angle K-N scattering (at more than "medium" energy, however) could provide nice QCD tests.

I will simply conclude by arguing that there is a good deal to be learned from high intensity sources of strangeness. I leave it to the reader to judge whether the levels of precision suggested by the "standard model" issues are attainable and/or desirable.

REFERENCES

1. D. Dicus and V. Mathur, Phys. Rev. D7, (1973)
   M. Veltman, Acta Phys. Pol. 138, 475 (1977) and Phys. Lett. 70B, 2531 (1973).
   B.W. Lee, C. Quigg and H.B. Thacker, Phys. Rev. Lett. 883 (1977) and Phys. Rev. D16, 1519 (1977).
2. S.L. Glashow, Nucl. Phys. 22, 579 (1961)
   S. Weinberg, Phys. Rev. Lett 19, 1254 (1967)
   A. Salam, Proc. 8th Nobel Symposium, ed. N. Svarthölm (Amqvist and Wiksell, Stockholm 1968) p. 367.
3. M.K. Gaillard and B.W. Lee, Phys. Rev. D10, 897 (1974).
4. S.L. Glashow, J. Iliopoulos and L. Maiani, Phys. Rev. D2, 1258 (1970).
5. M. Kobayashi and K. Maskawa, Prog. Theor. Phys. 49, 652 (1973).
6. For a more careful analysis, see J. Hagelin, these proceedings.
7. R.E. Shrock and M.B. Voloshin, Phys. Lett. 87B, 375 (1979).
8. T. Inami and C.S. Lim, Prog. Theo. Phys. 65, 297 (1981).
9. J. Ellis and J.S. Hagelin, Nucl. Phys. B217, 189 (1983).
10. T. Inami and C.S. Lim, Nucl. Phys. B207, 533 (1982).
11. A. Lahanas, private communication via J. Ellis.
12. For a review of these analyses see L.L. Chau, Brookhaven preprint BNL-31856 (1982), to be published in Phys. Reports.
13. An analysis based on recent data is K. Kleinknecht and B. Renk, Z. Physik C16, 7 (1982).
14. M.K. Gaillard, Y.-C. Kao, I.-H. Lee and M. Suzuki, Phys. Lett. 123B, 241 (1983).
15. F. Wilczek, these proceedings.
16. R.E. Shrock, Proc. 1982 DPF Summer Study on Elementary Particle

Physics and Future Facilities, Snowmass, Colorado, Eds. R. Donaldson, R. Gustafson and F. Paige, p. 291.
M. Suzuki, Berkeley preprint UCB-PTH-82/8.

17. J. Ellis, M.K. Gaillard, and D.V. Nanopoulos, Nucl. Phys. B109, 213 (1976).
18. F.J. Gilman and M.B. Wise, Phys. Letters 83B, 83 (1979). See also Ref. 12 for a review and extensive references.
19. A recently acclaimed method uses PCAC and SU(3) to relate $K^0 \to \bar{K}^0$ to $K^{\pm} \to \pi^{\pm}\pi^0$:
T. Appelquist, J.D. Bjorken and M. Chanowitz Phys. Rev. D7, 2225 (1973).
B.W. Lee, Proc. International Symposium on Lepton and Photon Interactions at High Energies, Ed. W.T. Kirk (SLAC, 1975) p. 635.
J.F. Donoghue et al. Phys. Lett. 119B, 412 (1982). While valuable as an independent estimate, I am not convinced that this is considerably more reliable than previous methods because of the approximations involved, in particular the retention of only linear terms in off-shell masses. The successful $K \to 3\pi$ analysis does not require a large off-shell extrapolation.
20. For a recent detailed analysis see F.J. Gilman and J.S. Hagelin, SLAC-PUB-3087 (1983).
21. T. Oka, Los Alamos preprint LA-UR-83-618 (1983).
22. P. Keller et al., Phys. Rev. Lett. 48, 971 (1982) and references therein.
23. M.K. Gaillard, Nuovo Cimento 6A, 559 (1971).
24. K. Kleinknecht, Proc. 17th International Conf. on High Energy Physics (London, 1974) p. III-23.
25. See the CP working group summary, these proceedings, for a detailed analysis of CP and T-violation in Λ and $\bar{\Lambda}$ decay. An early discussion of these effects within the context of the standard model is L.-L. Chau Wang, Proc. AIP Conf. "Weak Interactions as Probe of Unification" (VPI, 1980).
26. A.M.Diamant-Berger and R. Turlay, private communication (1976).
27. J. Ellis, M.K. Gaillard and D.V. Nanopoulos, Nucl. Phys. B106, 292 (1976).
28. N. Byers and J. Malenfant, in preparation.
29. B. Pire, private communication.

FAMILY SYMMETRIES

Frank Wilczek
Institute for Theoretical Physics, University of California
Santa Barbara, California 93106

ABSTRACT

Advantages accruing to theories with spontaneous breakdown of flavor symmetries are reviewed. A possible experimental signature in rare K decays is discussed. One particularly attractive breakdown scheme is pointed out, and its consequences for mixing angles described. Recent ideas concerning axions, and their possible cosmological significance, are briefly discussed. The relevance of Cavendish-type experiments in testing for axions and familons is mentioned.

§1 ADVANTAGES OF FAMILY SYMMETRIES

Although gauge theories have been remarkably successful in describing many features of the interactions of elementary particles, so far they have been rather disappointing in that little light has been shed on the pattern of fermion masses and mixing angles.

Perhaps the most outstanding feature of the fermion spectrum is the repetition of particles with the same SU(3) × SU(2) × U(1) quantum numbers: (e,μ,τ), (d,s,b), ... This pattern on its face suggests a symmetry of which these triplets are the multiplets. Postulating a continuous symmetry of this kind has several attractions:

i) The obvious one, that if the symmetry is broken in a simple pattern we may be able to relate observable fermion masses and mixing angles. This will be illustrated in §3.

ii) In order to understand the smallness of the CP-violating parameter θ in the string interactions, it seems that we need the Peccei-Quinn U(1) quasi-symmetry. This means that there must be field transformations which change the overall phase of the quark matrix, while leaving the rest of the Lagrangian unchanged. It is, I think, hard to believe that such a quasi-symmetry can be a fundamental physical law. Fortunately, a much more satisfactory possibility exists. The PQ quasi-symmetry can easily arise as an accidental by-product of more general genuine (spontaneously broken) family symmetries, and the usual requirements of renormalizability. As a toy example, let us consider a family symmetry SU(3) × SU(3) under which the mass matrix $M \to UMV$ $(U,V \in SU(3))$. Then the potential can contain terms like $\mathrm{tr}MM^{+}$, $\mathrm{tr}(MM^{+})^{2}$, $\mathrm{tr}(MM^{+})^{2}$, but not for instance $\mathrm{tr}M^{2}$, $\mathrm{tr}M^{4}$, ... The allowed terms do not depend on the phase of M (det M is forbidden by renormalizability), so there is an accidental PQ quasi-symmetry.

Along these lines, remember that the essence of the PQ mechanism consists in making the phase of the determinant of M a dynamical variable. It would then seem attractive a priori that more of M should be dynamical with the low-energy effective form determined by spontaneous symmetry breaking.

iii) The U(1) PQ quasi-symmetry in general contains a discrete subgroup of true symmetries. Spontaneous breakdown of these symmetries leads to the possibility of topologically stable domain walls, which are a grave cosmological embarrassment. If these discrete symmetries lie in a continuous group of family symmetries, the domain walls can relax away harmlessly.

§2 A POSSIBLE CONSEQUENCE: NAMBU-GOLDSTONE BOSONS

If there is a fundamental symmetry between the different families, broken not intrinsically but dynamically, then there will arise characteristic (strictly massless, spin zero, neutral) Nambu-Goldstone bosons. Some of these may be of phenomenological interest. The general form of coupling of these familons f is

$$\mathcal{L}_{\text{int.}} = \frac{1}{F}\,\partial^{\mu} f\, j_{\mu}$$

where F is the scale of symmetry breaking and j_{μ} is the relevant symmetry current, e.g. $j_{\mu} = \bar{s}\gamma_{\mu} d$ for $s \leftrightarrow d$ symmetry.

The decay $K^{+} \to \pi^{+} f$ looks most promising. One finds

$$\frac{K^{+}\to\pi^{+}f}{K^{+}\to\pi^{+}\pi^{0}} \approx 10^{14}\left(\frac{\text{GeV}}{F}\right)^{2}$$

The existing limit corresponds to $F \geq 10^{11}$ GeV. I will argue in §4 that $F \approx 10^{12}$ GeV is a very interesting value; it appears to be accessible in currently planned experiments.

Familons can also destabilize heavy neutrinos on cosmological time scales, circumventing the powerful limits ($m_{\nu} < 50$eV) on cosmologically stable neutrinos.

One might worry that truly massless particles might lead to large macroscopic effects, competing successfully with gravity. However the derivative coupling of familons suppresses these effects.

§3 A REPRESENTATIVE MASS MATRIX

I will now present an example of a family symmetry breaking that appears rather attractive.

Recall that in SU(5) fermion masses can be generated through a Higgs multiplet in the $\underline{5}$ representation. The resulting masses are equal in pairs between charge -1/3 quarks and charge -1 leptons, e.g. $m_b = m_{\tau}$. Actually equality only holds for effective masses at unification scales; to compare with laboratory results we must renormalize these down. This procedure gives us the good relation $m_b = 3m_\tau$ but also $m_e/m_\mu = m_d/m_s$ which is way off (m_d/m_s is calculated to be $\sim 1/20$ by current algebra methods).

Masses may also be generated by Higgs particles in the $\underline{45}$ representation; this contributes three times as much to lepton as to quark masses (before renormalization).

To first order the observed mass matrix is very simple: large entries for the third family, zeros elsewhere. It is simply accounted for if we group the fermion $\underline{5}$ and $\underline{10}$ representations into two triplets under a family SU(3), and assign the Higgs fields to (5,6) under (SU(5), SU(3)). The six-dimensional representation readily breaks giving a single non-zero component $\langle\phi_{33}\rangle \neq 0$; this yields the zeroth-order mass matrix including the good prediction for m_b/m_τ.

It is now entertaining to consider the hypothesis that the rest of the symmetry breaking is accomplished through a (45,3) Higgs representation. This will lead generically to a mass matrix of the form

$$\begin{pmatrix} 0 & A & 0 \\ -A & 0 & B \\ 0 & -B & C \end{pmatrix}$$

in the charge $-1/3$ quark sector. For the charged leptons, we get the same thing but with $B \to 3B$, $A \to 3A$ (for a Higgs $\underline{45}$) and an overall 1/3 (for renormalization effects).

The bad relationship $m_e/m_\mu = m_d/m_s$ gets replaced by the much better $m_e/m_\mu = 1/9\ m_d/m_s$. It is entertaining to suppose, in the absence of contrary evidence, that the mass matrix of the charge +2/3 quarks is diagonal. In this case we can readily infer the physical weak mixing angles given the form of the mass matrix in the charge $-1/3$ quark sector. One finds for the Cabibbo angle the famous successful relationship

$$\tan^2\theta_c = m_d/m_s$$

and for the analogous parameter governing the $b \to c$ transition

$$\tan^2\theta = m_s/m_b = \frac{1}{9}\, m_\mu/m_\tau$$

The latter formula is quite striking in that it implies very small angles and therefore a very long lifetime for the b-quark.

A feature which may be disturbing is that the bare strange-quark mass (renormalized at say 2GeV, before the QCD coupling has become really strong) is quite small, $\sim$35MeV. It is perhaps not excluded that m_s really is this small, although such crude estimates as exist in the literature are typically larger. The quantitative success of the quadratic Gell-Mann-Okubo mass formula for pseudoscalar mesons, which follows from treating m_s as a small parameter (presumably $m_s \ll f_\pi$, or $m_s \ll \Lambda_{QCD}$) may even be slight evidence for a tiny m_s.

§4 THE NEW! IMPROVED! AXION

Theoretical expectations for F, the scale at which the PQ quasi-symmetry is spontaneously broken, have varied over the years. The primary physical significance of F is that the mass and coupling strength of the axion are inversely proportional to F.

The original suggestion was $F \approx 300$ GeV, i.e. that the PQ quasi-symmetry breaks at the same scale as electroweak $SU(2) \times U(1)$. This suggestion was tested in several experiments and refuted; the axion was not found. Astrophysical arguments, based on how axion emission would affect the structure of stars, were used to bound $F \gtrsim 10^8$ GeV;

Later it was pointed out that in many unified models $F \approx 10^{15}$ GeV i.e. the PQ quasi-symmetry is broken at the same scale as the unification group (e.g. SU(5)). The axion is then exceedingly weakly coupled and therefore inaccessible to laboratory experiment. At this point it seemed we had a consistent, though fruitless, solution to the strong CP problem.

Very recently, however, a number of people realized that the invisible axion scheme has severe cosmological problems. There will typically be energy stored in the axion field, which of course contributes to the overall energy density of the Universe and is observable through its gravitational effects. It is easiest to visualize this process as the creation of a dense gas of axions, utterly cold, when the Universe is at $T \approx 1$ GeV. The mass density of the axion gas is calculated to be (to an adequate approximation) $10^4 \times (F/10^{15}$ GeV$)$ $\times\, \rho_B$, where ρ_B is the mass density of baryons. With $F \approx 10^{15}$ GeV, as for the invisible axion, this is too large.

An intriguing possibility is $F \approx 10^{12}$ GeV, which I cannot forbear calling the new! improved! axion. Then axions provide the non-baryonic, non-luminous dark matter for which impressive observational evidence now exists. Such a source of mass is also required for $\Omega = 1$, marginal closure, as is demanded by inflationary Universe ideas. Since the axions are produced cold they will behave like ultramassive (2 GeV) neutrinos as far as the theory of growth of inhomogeneities into galaxies is concerned. There is no "streaming" as there would be for light (≈ 30 eV) neutrinos. Thus axions probably lead to a hierarchical clustering picture instead of the Zeldovich fragmentation picture of galaxy formation. Recent work on galaxy formation in an axion-dominated Universe has been done by Shafi and Stecker,[1] Ipser and Sikivie,[2] and Turner, Zee, and me.[3] A dramatic development is the apparent observation of non-luminous matter clustering around dwarf galaxies.[4] It seems unlikely that ~ 30 eV neutrinos could have been gravitationally captured on such small scales, so something more exotic than a small neutrino mass would seem to be called for.

It seems very reasonable, in line with our arguments of §1 that PQ and family symmetries are closely related, that the scale $F \approx 10^{12}$ GeV should also be characteristic of family symmetry breaking generally. This idea can be tested in $K^+ \to \pi^+ f$, as mentioned in §2.

Finally with $F = 10^{12}$ GeV for family or PQ symmetries we might expect colored scalars with this sort of mass capable of mediating $\Delta B = 1$ decays. These tend to give proton decay at rates not incommensurable with existing experimental limits, and mostly into strange channels.

§5 WINDOWS ON FAMILY SYMMETRY

Let me briefly summarize the experimental handles on family symmetries discussed above. Potentially the richest of course is the pattern of fermion masses and mixing angles. A scheme such as that in §3 gives many relationships among these masses and angles as measured in weak and $\Delta B = 1$ decays, and also possible CP violation parameters. Of course observation of $K^+ \to \pi^+ f$ would be spectacular confirmation of this whole line of thought. Further cosmological investigations should determine whether we want the dark matter formed cold, as it will be if it is axions.

Another fascinating possibility is that very light axions or familons could be detected by macroscopic experiments. The Compton wavelength of the new! improved! axion is about 1 cm. Unlike familons it is not derivatively coupled and could contribute to Cavnedish-type experiments at short distances. Since $F^{-1} >> M_{p\ell}^{-1}$ the axions are in some sense more strongly coupled than gravitons, however their static coherent exchange violates CP and is thereby suppressed, though perhaps not hopelessly so. Both axions and familons could mediate spin-dependent $1/r^3$ forces, with a range of 1 cm or ∞ in the two cases.

Experimental input from direct observation of Higgs particles and their couplings would of course be most helpful. We must also be alive to the possibility, suggested in §4, that there can be substantial contributions to proton decay due to exchange of scalars.

Note: Much more detailed accounts of the material in §3 - §5 are being prepared, which will include adequate references. For §1 - §2, see Phys. Rev. Lett. 49, 1549 (1982).

ACKNOWLEDGMENT

This work was supported by the National Science Foundation, grant no. PHY77-27084.

REFERENCES

1. F. Stecker and Q. Shafi, Phys. Rev. Lett. 50, 928 (1983).
2. J. Ipser and P. Sikivie, Phys. Rev. Lett. 50, 925 (1983).
3. M. Turner, F. Wilczek, and A. Zee, Phys. Lett. 125B, 35 (1983).
4. M. Aaronson, Ap. J. 266, L11 (1983); S. Faber and D. Lin, Ap. J. 266, L17 (1983); D. Lin and S. Faber, Ap. J. 266, L21 (1983).

SUMMARY OF WORKSHOP ON CP-VIOLATION

Lincoln Wolfenstein and Darwin Chang
Physics Department, Pittsburgh, PA 15213

ABSTRACT

Observables showing CP-violation in K decays, Λ decay, and electric dipole moments are discussed. Possible values for these observables are discussed for three models: (1) Kobayashi-Maskawa model, (2) Weinberg's model with Higgs boson exchange and (3) $SU(2)_L \times SU(2)_R \times U(1)$ models.

1. CP-VIOLATING OBSERVABLES

So far CP-violation has been observed only in the K° system. Future experiments are designed either to increase the precision of our knowledge of CP-violation in the K° system or to find new systems for which CP-violation occurs. The superweak model makes precise predictions concerning the CP-violating parameters in K_L decay and at the same time predicts extremely small experimentally unobservable effects in most other systems. We now discuss a number of observables all of which are expected to be vanishingly small in the superweak model.

(1) ε'/ε. The observables in $K_L \to 2\pi$ are given in the standard form as

$$\eta_{+-} = \varepsilon + \varepsilon'/(1+\omega/\sqrt{2})$$

$$\eta_{oo} = \varepsilon - 2\varepsilon'/(1-\sqrt{2}\,\omega)$$

where $\omega = \mathrm{Re}A_2/\mathrm{Re}A_o \approx .05$. Because models are usually given with their own convenient phase convention it is best to give formulas[1] that are independent of the Wu-Yang convention

$$\sqrt{2}\,\varepsilon \approx e^{i\theta}\left(\frac{m'}{\Delta m} + \frac{\mathrm{Im}A_o}{\mathrm{Re}A_o}\right) \tag{1a}$$

$$\sqrt{2}\,\varepsilon' = e^{i\theta'}\left(\frac{\mathrm{Im}A_2}{\mathrm{Re}A_o} - \omega\,\frac{\mathrm{Im}A_o}{\mathrm{Re}A_o}\right) \tag{1b}$$

$$\theta \approx 44°,\ \theta' \approx 37 \pm 5°$$

where A_I is the decay amplitude for K° to a final ππ state with isospin I and m' is the imaginary part of the K° mass matrix. Here we have assumed CPT invariance which has the consequence that ε and ε' have almost the same phase so that for small ε'/ε we have

$$\left|\frac{\eta_{oo}}{\eta_{+-}}\right|^2 \approx 1 - 6(\varepsilon'/\varepsilon)$$

The measurement of this quantity is discussed in the talk of Winstein. An ongoing experiment E617 at FNAL should measure ε'/ε to .004; a proposal P731 should reduce the error to .001. Future experiments which Winstein could think of, which might be done at a kaon factory, reduce this error to $2\cdot10^{-4}$. As will be discussed later, exploring values of ε'/ε between 10^{-2} and 10^{-4} is extremely important.

The measurement of the phase of η_{+-} is primarily of interest in checking CPT. While this is by far the most precise test of CPT[2] there seemed no great interest in pushing this further. Another closely-related experiment is the charge asymmetry[3] in $K_L \to \pi\mu\nu$ or $K_L \to \pi e\nu$, which gives Re ε. It was pointed out that one can determine ε' from

$$\frac{\eta_{+-}}{(\text{Re }\varepsilon)\sec\theta} = 1 + \frac{\varepsilon'}{\varepsilon}$$

and that this determination at present competes with that from $|\eta_{oo}/\eta_{+-}|$. However it was felt that in the long run this was not the way to go after ε' and no great interest was expressed in pursuing the charge asymmetry measurements.

(2) $\underline{\eta_{+-o} - \eta_{+-}}$. This involves looking for CP-violation in the decay $K_s \to \pi^+\pi^-\pi^o$. Winstein mentioned the ongoing experiment of Thompson et al. at FNAL that hopes to measure this at a level of 10^{-3}. Such an accuracy will be difficult to achieve and it is hard to imagine improving on it. While a priori it is reasonable to expect $(\eta_{+-o} - \eta_{+-})$ to be of the order of $\varepsilon(\simeq 2.10^{-3})$ the models we discuss later give smaller values.

(3) Transverse polarization in $K \to \pi\mu\nu$. Schmidt from Yale discussed measurements of P_n, the correlation $(\vec{\sigma}_\mu \cdot \vec{p}_\pi x \vec{p}_\mu)$. Experiments carried out at Brookhaven give the results[4]

$$K_L^o \to \pi^-\mu^+\nu_\mu \qquad P_n^{lab} = 1.7 \pm 5.6 \times 10^{-3}$$

$$K^+ \to \pi^o\mu^+\nu_\mu \qquad P_n^{lab} = -3.7 \pm 4.7 \times 10^{-3}$$

In the absence of final state interactions a non-zero P_n is directly proportional to Imξ, which corresponds to a T-violating effect, where $\xi = f_-/f_+$, the ratio of the two form factors. For K_L^o

the electromagnetic interaction between π^- and μ^+ must be taken into account. This effect has been calculated to a high degree of accuracy as a function of the Dalitz plot;[5] it gives values of P_n^{lab} of the order 2 to 3 x 10^{-3}. Correcting for this the two experiments give $Im\xi = -.010 \pm .019$. Schmidt suggests that with a cleaner K_L beam, the present technique could yield an improvement of a factor of 10 and that he could imagine an improvement of a factor of 50.

(4) Λ decays. The decays $\Lambda \to p\pi^-$ and $\Lambda \to n\pi^0$ are usually analyzed in terms of the amplitudes $A_s(1)e^{i\delta_1}$, $A_p(1)e^{i\delta_{11}}$, $A_s(3)e^{i\delta_3}$, $A_p(3)e^{i\delta_{31}}$ where s(p) indicates s(p) waves in the final states and 1(3) indicates I = 1/2 (3/2) final states. The final state pion-nucleon phase shifts in these states are given by the indicated δ. In the absence of CP violation the A's are all real. Possible CP-violating phases θ_1, θ_3, ϕ may be defined by

$$\frac{A_s(1)}{A_p(1)} = \left|\frac{A_s(1)}{A_p(1)}\right| e^{i\theta_1} \qquad \frac{A_s(3)}{A_p(3)} = \left|\frac{A_s(3)}{A_p(3)}\right| e^{i\theta_3}$$

$$\frac{A_s(3)}{A_s(1)} = \left|\frac{A_s(3)}{A_s(1)}\right| e^{-i\phi} \tag{2}$$

We consider differences between Λ and $\bar{\Lambda}$ decays as a possible result of CP-violation.[6] Let the rate for $\Lambda \to \pi^- p$ be Γ and for $\bar{\Lambda} \to \pi^+ + \bar{p}$ be $\bar{\Gamma}$. Similarly let the asymmetry parameters for these decays be α and $\bar{\alpha}$. Then it is easy to show that to a reasonable approximation

$$\frac{\Gamma-\bar{\Gamma}}{\Gamma+\bar{\Gamma}} \approx \sqrt{2}\left|\frac{A_s(3)}{A_s(1)}\right| \sin\phi \sin(\delta_1-\delta_3) \approx 7\cdot 10^{-3} \sin\phi \tag{3a}$$

$$\frac{\alpha+\bar{\alpha}}{\alpha-\bar{\alpha}} \approx -\tan\theta_1 \tan(\delta_1-\delta_{11}) \approx -0.1 \tan\theta_1 \tag{3b}$$

where the last equalities involve insertion of experimental values. If the CP-violating phases are of the order of 10^{-3} then the asymmetry effect is of the order 10^{-4} and the rate effect is less than 10^{-5}. These are considerably lower than the estimates given by Chau.[7] As discussed below in specific models, it is highly probable that measurements of ε' provide a much more sensitive test of CP-violating decay amplitudes that do conceivable experiments comparing Λ and $\bar{\Lambda}$. No discussion of these experiments occurred at the workshop, but Kalogeropoulos[8] has estimated that the rate effect could be measured at the level of 10^{-4} at LAMPF II. As indicated above this is not sufficient accuracy. Further discussion of CP-violation in hyperon decays is given in the talk of Mary K. Gaillard.

(5) Rare K decays. In the talk of Gaillard there is a discussion of decays such as $K_L \to \pi^\circ e^+ e^-$, which are expected to be predominantly CP-violating. We did not explore these in this workshop. It is not clear what we would learn about CP violation if only the rate of this decay were measured.

(6) Electric dipole moments. A non-zero electric dipole moment of an elementary particle measures an effective P- and CP-violating diagonal interaction. Present limits on D_n, the electric dipole moment of the neutron, are $6 \cdot 10^{-25}$ e-cm from the Lobashov group.[9] Possibilities of reaching 10^{-26} or even lower have been frequently discussed. Limits on D_n in this range can constrain some models. There exists a possible CP-violating effect that goes by the name of the QCD θ, the value of which is very uncertain. From a phenomenological point of view there is only one appreciable effect of the QCD θ, which is a non-zero value for the electric dipole moment of hadrons. Thus a non-zero value of D_n could always be blamed on θ and it would tell us nothing about models of CP-violation in K decay or other processes. This means that models can be ruled out if experimentally D_n is too low but not if D_n is too large. Gavela emphasized the possible significance[10] of measuring the electric dipole moment of the electron D_e, which can be done in atomic physics experiments. Unlike the case of D_n, a non-zero D_e cannot be blamed on the QCD θ. However, many models that predict an interesting value of D_n predict a very small value for D_e.

2. MODELS

The models we will discuss are all of the milliweak type in the sense that the effective Hamiltonian has a CP-conserving piece and a CP-violating piece with the latter having a coefficient of the order of 10^{-3}. These are to be contrasted to the superweak models in which the CP-violating piece has a coefficient of the order of 10^{-9}, which is significant only because it allows $\Delta S = 2$ and thus can contribute to the K°-$\overline{K}^\circ$ mass matrix at tree level. We have defined the observables in the last section so that they all are expected to be negligibly small for superweak models. The name of the game is to find deviations from superweak that may serve as clues to the correct milliweak model. So far none have been found. It should also be emphasized that there are many possible gauge models that are superweak. When one goes from the K system to heavy quark systems like D and B, these models need not have the same prediction. In particular, in superweak Higgs models the CP-violating effects may be proportional to the quark mass.

It is simplest to think in terms of competing alternative models. However it may be more correct to think of various different mechanisms of CP-violation all of which might coexist. While one mechanism may well dominate in $K_L \to 2\pi$ decay, another mechanism may be more important for some other observables. (We have already noted this for the case of D_n.) Three models were discussed in detail in our workshop. They serve as interesting guides to the sensitivity of different experiments. Their predictions are summarized in Table 1. It should be emphasized, however, that none of these models is very convincing; none of them yields the gut reaction: "Now I understand CP-violation". The most probable explanation of CP-violation thus comes under the heading "other" for which we give no predictions.

(1) KM model. Within the framework of the minimal Weinberg-Salam model the only source of CP-violation is the Kobayashi-Maskawa (KM) mechanism.[11] This involves including a complex phase in the quark mass matrix. One may consider this model as blaming our failure to understand CP-violation as part of our failure to understand masses; thus two mysteries are compressed into one. If there are only six quarks there is only one phase and therefore only one CP-violating parameter. If the other elements of the mass matrix were known (such as the t-quark mass and mixing angles θ_2 and θ_3), as we hope they eventually will, then all CP-violating effects could in principle be predicted from the knowledge of one. Efforts to do this are severely impeded by the problem of calculating hadron dynamics.

In 1979 Gilman and Wise pointed out that ε'/ε was not negligibly small in the KM model.[12] With the standard phase convention, CP-violation occurs only for the $\Delta I = 1/2$ amplitude A_o so that Eq. (1b) gives

$$\sqrt{2}\ \varepsilon' = -e^{i\theta'}\ (.05)\ \mathrm{Im}A_o/\mathrm{Re}A_o \qquad (4)$$

They pointed out that $\mathrm{Im}A_o$ had a significant contribution from penguin graphs, which are believed to make a major contribution to $K^\circ \to 2\pi$ decay. Their calculations based on assumed values for m_t and the mixing angles and optimistic dynamic assumptions gave $\varepsilon'/\varepsilon = .01$ to .02. Guberina and Peccei using the same physics but a different way of treating the dynamics found a value around $3\cdot10^{-3}$. Wolfenstein, following the method of Gilman and Wise, but allowing for the uncertainty in the dispersive contributions to the mass matrix and varying the mixing angles, found a minimum value of $|\varepsilon'/\varepsilon|$ of about $5\cdot10^{-3}$. Recently Gilman and Hagelin have used the data on $K_L \to \mu^+\mu^-$ to constrain the parameters of the mass matrix; then following the methods of Guberina and Peccei they have found a minimum value of $|\varepsilon'/\varepsilon|$ of $2\cdot10^{-3}$. Due to the uncertainty in evaluating matrix elements the number 2 is uncertain by at least a factor 2. The sign of ε'/ε is predicted to be positive. These results show that it is very important to measure ε'/ε to an accuracy of a part per thousand. If the KM model is correct it is expected that a non-zero value will then be found.

From the summary in Table 1 there are no other attractive experiments in the case of the KM model. We comment only on the new results given to this conference by Chang for the case of Λ decay. For all $\Delta S = 1$ decay amplitudes governed by penguins one has that ImA/ReA has the same values as ImA_0/ReA_0 in $K^\circ \to 2\pi$ decay. $\Delta S = 1$ decay amplitudes due to simple tree graphs have no CP-violation with the standard KM phase convention. Thus it is possible to relate all CP-violating $\Delta S = 1$ decay amplitudes to ε' from Eq. (4). The result for Λ decay using Eqs. (3) is

$$\left|\frac{\Gamma-\bar{\Gamma}}{\Gamma+\bar{\Gamma}}\right| = 0.2\ |\varepsilon'|\ f_s \qquad (5a)$$

$$\left|\frac{\alpha+\bar{\alpha}}{\alpha-\bar{\alpha}}\right| = 3\ |\varepsilon'|\ (f_p-f_s) \qquad (5b)$$

where $f_s(f_p)$ is the fraction of $\Delta I = 1/2$ s-wave (p-wave) Λ decay due to penguins compared to the fraction for $K^\circ \to 2\pi$. On the basis of the discussion of Donoghue et al.[13] we expect that f_s is considerably smaller than unity while f_p might be close to unity. In particular for ε'/ε of the order 10^{-2} to 10^{-3} we obtain values of $(\Gamma-\bar{\Gamma}/\Gamma+\bar{\Gamma})$ of the order 10^{-6} to 10^{-7}, much too small to be of interest.

(2) Weinberg Higgs model.[14] In the context of four quarks Weinberg suggested that the minimal SU(2) x U(1) model should be extended to contain three Higgs doublets in order to allow CP-violation. Branco pointed out that if the CP-violation were spontaneous then even with six or more quarks the KM phase would be zero so that Higgs exchange would be the only source of CP violation. It was pointed out by Deshpande and independently by Sanda that as a result of Higgs penguin diagrams one expects ImA_0/ReA_0 to be much larger than $m'/\Delta m$ so that from Eqs. (1)

$$\varepsilon'/\varepsilon = -\omega = -.05$$

a result practically ruled out by existing experiments. (It was also noted at this conference by Hagelin that the sign was opposite to that of the KM model. The reason, looking at Eqs. (1), is that for the KM model $m'/\Delta m$ is larger than ImA_0/ReA_0 and of opposite sign). It was pointed out by Chang[15] and later also by Dupont and Pham and by Hagelin[16] that the inclusion of dispersive terms in the mass matrix could reduce this result by a factor of as much as 3. In any case the present round of experiments must soon find a non-zero value of ε'/ε or this model is ruled out.

Of the other predictions for this model listed in Table 1, the tranverse muon polarization in $K \to \pi\mu\nu$ (which measures $Im\xi$) is of particular interest. This parameter is associated with an effective scalar exchange and so may be expected to be significant in a model with Higgs exchange. Calculation[17] yields in our notation[14,15]

$$\mathrm{Im}\xi = \frac{\sqrt{2}}{4}\,\frac{m_K^2}{m_o^2}\left(\frac{v_1}{v_3}\right)^2$$

Here m_o is a mass scale related to the Higgs mass and so defined that CP-violation in $K \rightarrow 2\pi$ depends mainly on m_o. In particular

$$m'_{box}/\Delta m \simeq 2\cdot 10^{-3}\left(\frac{2\ \mathrm{GeV}}{m_o}\right)^2$$

This yields the estimate $m_o \sim 2$ GeV if $\mathrm{Im}A_o/\mathrm{Re}A_o$ is small, but as a result of the Higgs penguin contributions to $\mathrm{Im}A_o/\mathrm{Re}A_o$ we expect that, $m'_{box}/\Delta m$ is much smaller than ε so that m_o is more like 6 to 10 GeV. The quantities v_1^{-1} and v_3^{-1} are proportional to the couplings of two of the original Higgs fields ϕ_1 and ϕ_3 where ϕ_1 couples to down quarks and it is assumed ϕ_3 couples to the leptons. To get a result for $\mathrm{Im}\xi$ it is necessary to assume these have the same order of magnitude; one then gets the order-of-magnitude result

$$\mathrm{Im}\xi \sim 10^{-3}$$

The result we quote is lower than that of Ref. 17 because of the inclusion of the Higgs penguin graphs.

It was pointed out by several discussants that even if the major contribution to CP-violation in $K^\circ \rightarrow 2\pi$ decay were not due to Higgs exchanges, one should still consider this mechanism. The reason lies in the fact that many models (such as technicolor) involve the existence of extra Higgs bosons, which in general could mediate CP-violating effects. Even if the Higgs boson were so massive that it made only a small contribution ($\delta\varepsilon$) to ε in $K^\circ \rightarrow 2\pi$ decay it might still dominate other CP-violating parameters as $\mathrm{Im}\xi$ in $K \rightarrow \pi\mu\nu$. In fact if the Higgs made only a 5% contribution to ε it could still give a sizeable contribution ($\sim 2.5 \times 10^{-3}$) to ε'/ε. (If the Higgs contribution were added to the KM contribution the two could conceivably, but not necessarily, cancel each other.) The estimates given in Table 1 should all be scaled down by the factor $(\delta\varepsilon/\varepsilon)$ in this case.

(3) <u>$SU(2)_L \times SU(2)_R \times U(1)_{B-L}$ models</u>. This represents one of the simplest and most attractive extensions of the $SU(2)_L \times U(1)$ gauge group. Such models allow for the possibility of a left-right (LR) symmetry in the Lagrangian that is broken spontaneously. The quarks in this model are (i = generation index)

$$Q_{Li} = \begin{pmatrix} u_L \\ d_L \end{pmatrix}_i \qquad Q_{Ri} = \begin{pmatrix} u_R \\ d_R \end{pmatrix}_i$$

transforming as $(2, 1, \frac{1}{3})$ and $(1, 2, \frac{1}{3})$, respectively. The Higgs bosons that give fermion masses transform as (2, 2, 0)

$$\phi = \begin{pmatrix} \phi_1^{\circ} & \phi_1^{+} \\ \phi_2^{-} & \phi_2^{\circ} \end{pmatrix} \rightarrow \begin{pmatrix} z & 0 \\ 0 & z' \end{pmatrix} \tag{6}$$

where the arrow indicates vacuum expectation values (vev). To break the left-right symmetry one may introduce Δ_L (3, 1, 2) and Δ_R (1, 3, 2) where Δ_R developes a large vev. The LR symmetry of the Lagrangian corresponds to the discrete symmetry

$$Q_L \leftrightarrow Q_R, \ \phi \leftrightarrow \phi^+ \quad \Delta_L \leftrightarrow \Delta_R$$

The Yukawa coupling can be written

$$\mathcal{L}_y = \overline{Q}_L \hat{f} \phi Q_R + \overline{Q}_L \hat{h} \tilde{\phi} Q_R + \text{h.c.}$$

where $\hat{f}$ and $\hat{h}$ are 3 x 3 hermitean matrices in the generation space. The mass matrices for the quarks are given by

$$\begin{aligned} \hat{M}^u &= \hat{f}\, z + \hat{h}\, z'^* \\ \hat{M}^d &= \hat{h}\, z^* + \hat{f}\, z' \end{aligned} \tag{7}$$

The diagonalization of these matrices leads in general to a mixing matrix (KM matrix) U_L for the exchange of W_L which can be different from U_R for exchange of W_R. For the case we consider in which $\hat{f}$ and $\hat{h}$ are real, the mixing angles are the same and the only difference is associated with the phases. There are N(N+1)/2 unremoveable phases for the N generation case, thus 3 for N = 2 and 6 for N = 3.[18] In addition to these extra phases the model has two new parameters

$$\begin{aligned} r &= |z'/z| \\ \beta &= |m(W_L)/m(W_R)|^2 \end{aligned} \tag{8}$$

Since βr determines the mixing of W_L and W_R it is tightly constrained by experiments.[19]

It was first pointed out by Mohapatra and Pati[20] in the two generation case that the extra phases could lead to CP-violation due to the exchange of W_R even though there was none due to exchange of W_L alone. Assuming W_L-W_R mixing was negligible they showed that there exists a symmetry, called the "isoconjugate relation" that insures $\varepsilon' = 0$. They also assert that $|\eta_{+-o} - \eta_{+-}|$ might be of the order of ε in this model. When this model is extended to six quarks one expects in general there will be a CP-violating phase (KM phase) associated with W_L exchange alone and so that model becomes the KM model with some modifications due to W_R exchange. Branco et al.[21] have shown that it is possible to impose further symmetries to eliminate the complex phase in the W_L exchange and so make ε'/ε zero or very small.

From here on we shall assume that CP-invariance is spontaneously broken. In this case $\hat{f}$ and $\hat{h}$ are real so that $\hat{M}^u$ and $\hat{M}^d$ are symmetric (but complex). This implies that there exists a phase convention such that $U_L = U_R^*$. This case may be called "pseudo-manifest" L-R symmetry. The Higgs structure discussed above leads to a minimal model of spontaneous CP-violation recently given by Chang.[22] Because of the tight constraints of this model it is possible to make more specific predictions than in the general case. Without loss of generality we set in Eq. (6) $z = K$ and $z' = K'e^{i\alpha}$ with K and K' positive and real. It follows that all CP violating observables must be proportional to the parameter $\frac{K'\sin\alpha}{K} \equiv r\sin\alpha$. Since r determines the W_L-W_R mixing, there is a direct relation between CP-violation and W_L-W_R mixing. Since there is only one CP-violating parameter it is possible to calculate the CP-violating phase (KM phase) associated with W_L exchange in terms of this parameter; this explicit calculation shows that CP-violation associated with W_L exchange is relatively unimportant. As a result there is no qualitative change in going from four quarks to six, although there are then more uncertainties associated with mixing angles and m_t. The results for the observables given in Table 1 and discussed below were calculated for the four-quark case. In doing the calculations the phase convention is employed in which U_L is real and the three CP-violating phases occur in U_R.

For the calculation of ε'/ε one first looks at m' which comes from a box diagram with one W_L and one W_R, yielding

$$m'/\Delta m \simeq -\frac{430}{2}\left(\frac{m_c}{m_s}\right) \beta r \sin\alpha \tag{9}$$

The large factor 430 was first noted by Beall et al.[23] in their calculation of the contribution of the W_L-W_R box to Δm. (On this basis they suggested that to avoid getting too large a value for Δm one needs $\beta \lesssim (1/430)$; that is $m(W_R) \gtrsim 20\, m(W_L)$. Possibilities of evading this requirement in the six-quark case were presented to the workshop by Senjanovic.[24]) Because of this large factor, it turns out that $\mathrm{Im}A_o/\mathrm{Re}A_o \ll m'/\Delta m$ so that from Eq. (1a)

$$\sqrt{2}\,\varepsilon\, e^{-i\theta} = -\frac{1}{2}(430)\frac{m_c}{m_s}\beta r \sin\alpha \tag{10}$$

The CP-violating decay amplitudes come from Fig. (1a), (1b) with $W = W_R$ and Fig. (1c). If we ignore Fig. (1c) we get the Mohapatra-Pati result that $\varepsilon' = 0$ since CP-violation is just an overall relative phase between Fig. (1a), (1b) with W_R exchange and the same diagrams with W_L exchange. However, as noted above, in the Chang model all CP-violation is proportional to the mixing parameter r and so Fig. (1c) cannot be ignored. One obtains

$$\sqrt{2}\,\varepsilon'\, e^{-i\theta'} = 4\omega\beta r\,(\sin\alpha)\,\chi$$

$$\chi \equiv \frac{(A_o)_{P'}}{(A_o)_P} - \frac{(A_2)_{P'}}{(A_2)_P} \tag{11}$$

where the subscript P' designates the normalized contribution from Fig. (1c) whereas P is that from Figs. (1a), (1b). Taking $m_c/m_s \gtrsim 10$

$$\frac{\varepsilon'}{\varepsilon} \lesssim 10^{-4}\chi \tag{12}$$

The contributions of Fig. (1c) are in fact enhanced because they involve an L-R operator; this is particularly true of A_2 because of the short-distance correction[19]. A rough estimate for χ is between 10 and 100 yielding

$$|\varepsilon'/\varepsilon| = 10^{-2} \text{ to } 10^{-3}$$

The result overlaps very much the prediction of the KM model and again emphasizes the importance of measurements at the accuracy of a part in a thousand.

The value of $\eta_{+-o}-\eta_{+-}$ is determined by the relative phase between the parity-even and parity-odd parts of the Hamiltonian. This is found to be

$$\eta_{+-o} - \eta_{+-} = 2i\,\frac{m_c}{m_s}\,\beta r \sin\alpha$$

$$\left|\frac{\eta_{+-o}-\eta_{+-}}{\eta_{+-}}\right| = \frac{4\sqrt{2}}{430} \sim 10^{-2}$$

a discouraging result. Note it is the factor 430 that depresses the result below unity. Thus if one accepted the special case discussed by Senjanovic[24] this result would need to be re-evaluated, as would the other observables. No detailed calculations have been made for Λ decay, but as in the KM model we expect that the dominant diagrams that contribute to the CP-violating amplitudes A for Λ decay are the same that contribute in $K^\circ \to 2\pi$. Therefore we expect qualitatively similar results as in the KM model.

The neutron electric dipole moment D_n has been calculated for L-R models in general by Beall and Soni.[25] Applied to the minimal Chang model this gives

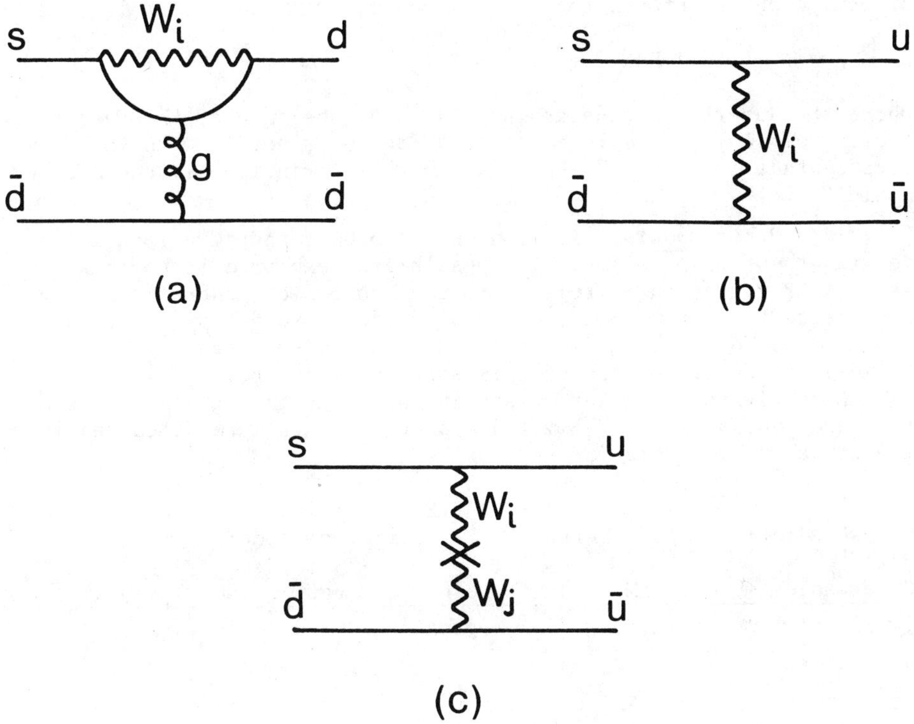

Fig. 1. Diagrams contributing to $K^{\circ} \rightarrow 2\pi$ decay. W_i can be either W_L or W_R. In (c) $i \neq j$ corresponding to $W_L - W_R$ mixing. g stands for gluon. Solid lines are quark lines.

$$D_n \lesssim (5.6 \times 10^{-22}\ \text{e-cm})\ 4\beta r \sin\alpha$$
$$\lesssim 3 \times 10^{-27}\ \text{e-cm}$$

which is much larger than the KM model and may be accessible to future experiments.

Finally it may be noted[26] that in models with W_L-W_R mixing there is a non-negligible value for the T-violating D parameter in beta-decay such as that of the neutron or Ne^{19}. This arises from an effective phase between the V and A couplings. The result is that

$$D \simeq 2\beta r \sin\alpha \lesssim 3\cdot 10^{-6}$$

where the numerical value again is so small because of the factor 430 in Eq. (10). This T-violating effect does not show up in $K \rightarrow \pi\mu\nu$ because that is a pure V decay, although in principle it could show up in K_{e4}.

(4) Other models. It is natural to hope that CP-violation is related to other problems. One possibility explored in various ways is that CP violation is related to the problem of generations. In some sense this is the case of the KM model. More explicitly this can be done in models in which the horizontal symmetry is gauged. No very attractive models of this sort have appeared.

Recently interest has arisen in the relation between CP-violation and supersymmetry. Limited aspects of this were discussed at the workshop by Gavela, and by Bigi.

TABLE I

Predictions for observables for different models.

Observable	KM	Weinberg Higgs	$SU(2)_L \times SU(2)_R$
$\lvert\varepsilon'/\varepsilon\rvert$	$\gtrsim 2\cdot 10^{-3}$	$\gtrsim 2\cdot 10^{-2}$	10^{-2} to 10^{-3}
$\dfrac{\eta_{+-0}-\eta_{+-}}{\eta_{+-}}$	$\sim\varepsilon'/\varepsilon$		$\sim 10^{-2}$
$K \rightarrow \pi\mu\nu : \text{Im}\xi$	0	$\sim 10^{-3}$	0
$\dfrac{\Gamma(\Lambda\rightarrow p)-\Gamma(\bar{\Lambda}\rightarrow\bar{p})}{\text{Sum}}$	$\lesssim\varepsilon'/5$		$\sim\varepsilon'$
$\dfrac{\alpha(\Lambda\rightarrow p)+\bar{\alpha}(\bar{\Lambda}\rightarrow\bar{p})}{\text{Difference}}$	$\sim 3\varepsilon'$		$\sim\varepsilon'$
D_n	$<10^{-30}$	$\sim 5\cdot 10^{-26}$	$\sim 3\cdot 10^{-27}$

REFERENCES

1. See L. Wolfenstein in Theory and Phenomenology in Particle Physics, ed. A. Zichichi (Acad. Press 1969), p. 218. Our ε' is obtained by subtracting $\sqrt{\frac{1}{2}}\,\varepsilon\omega$ from Eq. (18b) for ε' in this reference.
2. See L. Wolfenstein, Nuov. Cim. 63A, 269 (1969).
3. For an experimental review of these and other observables, see K. Kleinknecht, Ann. Rev. Nuc. Sci. 26, 1 (1976).
4. M. K. Campbell, et al., Phys. Rev. Lett. 47, 1032 (1981); Phys. Rev. D21, 1750 (1980); Phys. Rev. D27, 1056 (1983).
5. L. B. Okun and I. B. Khriplovich, Phys. Lett. 24B, 672 (1967) and G. S. Adkins, Princeton preprint give explicit formulae for constant form factors. The small effect of q^2-dependence in the form factors is illustrated by J. Brodine, Nuc. Phys. B30, 545 (1971) and Carnegie-Mellon Thesis (1970). The very small final state interaction effect in K^+ decay is estimated in Ref. 17.
6. O. Overseth and S. Pakvasa, Phys. Rev. 184, 1663 (1969).
7. L.-L. Chau, BNL Report 31859-R (to be published in Phys. Reports).
8. T. Kalogeropoulos in "Proceedings of the Second LAMPF II Workshop", H.A. Thiessen et al., editors (LANL internal pub. 1982) p. 617.
9. I. S. Altarev, et al., Phys. Lett. 102B, 13 (1981).
10. M. B. Gavela and H. Georgi, Phys. Lett. 119B, 141 (1982).
11. For a review see Ref. 7.
12. F. Gilman and M. B. Wise, Phys. Lett. 83B, 83 (1979), Phys. Rev. D20, 2392 (1979). For an earlier review of this work see L. Wolfenstein in Particles and Fields, AIP Conference Proceedings No. 59, p. 365. The recent work is reviewed in the contribution of J. Hagelin to these proceedings. The parameter called ξ by Hagelin is equal to ImA_0/ReA_0.
13. J. F. Donoghue et al., Phys. Rev. D21, 186 (1980), Tables 3 and 5.
14. This is reviewed in the contribution of N. Deshpande. See references therein.
15. D. Chang, Phys. Rev. D25, 1318 (1982).
16. Y. Dupont and T. N. Pham, CNRS report A498-0482 (1982); J. Hagelin, Phys. Lett. 117B, 441 (1982).
17. A. Zhitnitskii, Sov. J. Nuc. Phys. 31, 529 (1980).
18. Without L-R symmetry the number is N^2-N+1; the result of L-R symmetry is to reduce the number by one for N=3.
19. See, for example, I. I. Bigi and J. M. Frere, Phys. Lett. 110B, 255 (1982); J. Donoghue and B. Holstein, Phys. Lett. 113B, 382 (1982).
20. R. N. Mohapatra and J. C. Pati, Phys. Rev. D11, 566 (1975).
21. G. C. Branco, J. M. Frere, and J. M. Gerard, CERN preprint TH3406; G. C. Branco and J. M. Frere, contribution to these proceedings.
22. D. Chang, Nuc. Phys. B214, 435 (1983).
23. G. Beall, et al., Phys. Rev. Lett. 48, 848 (1982).

24. R. N. Mohapatra, et al., Brookhaven preprint; G. Senjanovic, contribution to these proceedings.
25. G. Beall and A. Soni, Phys. Rev. Lett. 47, 552 (1981).
26. This effect was first noted for L-R models with W_L-W_R mixing by P. Herczeg, Phys. Rev., to be published.

A LOWER BOUND ON $|\epsilon'/\epsilon|$

JOHN S. HAGELIN AND FREDERICK J. GILMAN*
Stanford Linear Accelerator Center
Stanford University, Stanford, California 94305

ABSTRACT

In the Kobayashi-Maskawa (K-M) model, a lower bound on the CP violating product $s_2c_2s_3s_\delta$ follows from the imaginary part of the short-distance $K^0-\bar{K}^0$ mixing amplitude together with a conservative upper bound on the short-distance contribution to $K_L \to \mu\mu$. This leads to a lower bound on $|\epsilon'/\epsilon|$ in terms of a matrix element of a single $(V-A)\times(V+A)$ type operator. Familiar current algebra and bag model estimates for this matrix element give $|\epsilon'/\epsilon| > 2\times10^{-3}$. We also observe that the experimental upper bound on the branching ratio for the b-quark into u-quarks fixes the sign of $s_2c_2s_3s_\delta$ and ϵ'/ϵ both to be positive. Allowances for QCD corrections and long distance effects are included throughout our analysis.

There are in general two possible sources which contribute to the observed non-invariance of CP in the neutral kaon system[1]: (1) "direct" CP violation in the $\Delta S = 1$ Hamiltonian responsible for $K \to 2\pi$ decay, and (2) a "superweak" contribution which is purely $\Delta S = 2$, contributing to CP violation in the $K^0-\bar{K}^0$ transition amplitude. Direct CP violation in $K \to 2\pi$ can give rise to different phases for the weak amplitudes for $K \to 2\pi\ (I=0)$ and $K \to 2\pi\ (I=2)$ decays, and is parameterized by the quantity ϵ' where

$$\epsilon' = \frac{1}{\sqrt{2}}\ exp\left[i(\frac{\pi}{2}+\delta_2-\delta_0)\right]\frac{ImA_2}{A_0}\ , \tag{1}$$

in the standard phase convention where A_0 is chosen to be real and positive.[2] Superweak CP violation cannot produce different phases for A_0 and A_2, but contributes to the CP impurity parameter ϵ which measures the departure of the mass eigenstate K_L, K_S from CP eigenstates:

$$\epsilon = \frac{(\frac{1}{2}Im\Gamma_{12}+iImM_{12})}{(\Delta M-\frac{1}{2}\ i\Delta\Gamma)}\ , \tag{2}$$

* Work supported by the Department of Energy, contract DE-AC03-76SF00515.

where M_{12} and Γ_{12} are the off-diagonal elements of the $K^0 - \bar{K}^0$ mass and decay matrices.

In the standard convention where A_0 is real, $Im\Gamma_{12}$ is negligible [resulting from the fact that the 2π $(I = 0)$ contribution dominates the width difference] which together with the approximate relation $\Delta M = -\frac{1}{2}\Delta\Gamma$ allows one to write

$$\epsilon = \frac{exp\,(i\pi/4)}{2\sqrt{2}} \left(\frac{ImM_{12}}{ReM_{12}}\right)^{A_0^R} \quad , \tag{3}$$

where the superscript A_0^R is intended to emphasize that the quantities are to be evaluated in the convention where A_0 is real and positive.

In general, the conventional choice of quark phases in which the weak couplings to light quarks are real is not consistent with the convention in which A_0 is real, unless the $\Delta S = 1$ Hamiltonian responsible for $K \to 2\pi$ $(I = 0)$ decay receives no CP violating contribution from loop diagrams. We know, however, that in the Kobayashi-Maskawa (KM)[3] model, "penguin" diagrams generate an effective $\Delta S = 1$ interaction which is CP violating.[4] To the extent that such "direct" CP violation is confined to penguin-type operators, it is purely $\Delta I = \frac{1}{2}$ and contributes only to A_0:

$$A_0^Q = exp\,(i\xi)|A_0^Q| \quad , \tag{4}$$

where here the superscript Q is used to emphasize a quark convention in which A_0 is complex.

The standard phase convention where A_0 is real is restored simply by redefining the phases of the K^0 and $\bar{K}^0$ states: $|K^0> \to e^{-i\xi}\,|K^0>$, $|\bar{K}^0> \to e^{i\xi}\,|\bar{K}^0>$, so that $A_0^Q \to e^{-i\xi}A_0^Q = A_0^R$. At the same time, the previously (in the quark basis) real amplitude A_2 picks up a phase $e^{-i\xi}$ and is complex in the basis where A_0 is real. Thus from Eq. (1),

$$\left|\frac{\epsilon'}{\epsilon}\right| = \frac{1}{\sqrt{2}}\frac{|\xi|}{|\epsilon|}\left|\frac{A_2}{A_0}\right| = 15.6|\xi| \quad , \tag{5}$$

where we have used the experimental values[5] of $|A_2/A_0| = 1/20$ and of $|\epsilon| = 2.27 \times 10^{-3}$.

The effective $\Delta S = 1$ Hamiltonian which is responsible for K^0 decay has been extensively studied elsewhere.[6] The CP-violating contribution to $K^0 \to \pi\pi$ $(I = 0)$ decay is dominated by the contribution from a single $(V - A) \times (V + A)$ operator, Q_6, in the effective Hamiltonian $\mathcal{H} = \sum_{i=1}^{6} C_i Q_i$. ImC_6 is proportional to the combination of K-M parameters $s_2 c_2 s_3 s_\delta$, in addition to the usual factor of $\frac{G_F}{\sqrt{2}} s_1$ characteristic of $\Delta S = 1$ weak amplitudes. Thus we write

$$\begin{aligned}\xi &= \frac{Im < \pi\pi(I=0)|\mathcal{H}|K^0 >}{A_0} \\ &\approx \frac{Im C_6 < \pi\pi(I=0)|Q_6|K^0 >}{A_0} \qquad (6) \\ &\equiv (s_2 c_2 s_3 s_\delta)(Im\, \tilde{C}_6) \frac{G_F}{\sqrt{2}} s_1 \frac{< \pi\pi(I=0)|Q_6|K^0 >}{A_0} ,\end{aligned}$$

where $G_F s_1/\sqrt{2}$ and A_0, the $K_0 \to \pi\pi$ $(I=0)$ amplitude, have values directly determined by experiment, which we will use. $Im\, \tilde{C}_6$ and $< \pi\pi(I=0)|Q_6|K^0 >$ have been studied and discussed elsewhere, and to these we shall return. Our objective is to establish a lower bound on the CP violating product $s_2 c_2 s_3 s_\delta$ to which we now proceed.

To constrain the product $s_2 c_2 s_3 s_\delta$ we return to the expression for ϵ in Eq. (3) in the basis where A_0 is real. Here much previous work used a short distance analysis[7] for both ReM_{12} and ImM_{12}. However it is difficult to justify neglecting the long distance contributions to ReM_{12}, and we shall simply use the experimental value for $\Delta M = 2ReM_{12}$ in the denominator of Eq. (3). On the other hand, one can argue that ImM_{12} is given almost entirely by the short distance contribution in the phase convention where A_0 is real.[8] Traditionally, this short-distance (sd) contribution (from the box diagram involving heavy quarks and W's) is computed in the quark basis. In the basis where A_0 is real, $ImM_{12}^{sd} = (ImM_{12}^{sd})^Q + 2\xi ReM_{12}^{sd}$, and Eq. (3) becomes

$$\epsilon = \frac{e^{i\pi/4}}{\sqrt{2}} \left[\frac{(ImM_{12}^{sd})^Q}{\Delta M} + 2\xi \frac{ReM_{12}^{sd}}{\Delta M} \right] \qquad (7a)$$

$$= \left[\frac{BG_F^2 f_K^2 m_K}{12\sqrt{2}\,\pi^2 \Delta M_K} Im\left(\eta_1 \lambda_c^2 m_c^2 + \eta_2 \lambda_t^2 m_t^2 + 2\eta_3 \lambda_c \lambda_t m_c^2 \ell n \frac{m_t^2}{m_c^2} \right) + \sqrt{2}\,\xi \frac{ReM_{12}^{sd}}{\Delta M} \right] e^{i\pi/4} \qquad (7b)$$

where $\lambda_q \equiv U_{qs}^* U_{qd}$ is a product of K-M matrix elements, B parameterizes the matrix element of the $\Delta S = 2$ operator ($B = +1$ for vacuum insertion),[7] and η_1, η_2, η_3 take account of the strong interaction corrections to the effective $\Delta S = 2$ Hamiltonian relevant to $K^0 = \bar{K}^0$ mixing, and take the values 0.7, 0.6 and 0.4, respectively, for $M_W = 80$ GeV and $\Lambda_{QCD} = 0.1$ GeV.[9,10] Although not written expressly in Eq. (7), we also include the effect[11,12] of higher powers in

m_t^2/M_W^2. The value of B has recently been extracted by relating the matrix element of the $\Delta S = 2$ operator to the measured contribution of the $\Delta I = 3/2$ operator to $K \to \pi\pi$ decay using $SU(3)$ and current algebra, giving[13] $B \simeq 0.33$. Inserting experimental masses and dropping terms which are second order in $sin\theta_i$ compared to those of zeroth order, Eq. (7b) becomes

$$\epsilon = \left[\left(\frac{1.94}{GeV^2}\right) s_1^2(s_2c_2s_3s_\delta)\left(-0.7m_c^2 + 0.6m_t^2\left(\frac{Re\lambda_t}{s_1}\right) + 0.4m_c^2\ell n\left(\frac{m_t^2}{m_c^2}\right)\right)\right.$$
$$\left. + \sqrt{2}\,\xi\frac{ReM_{12}^{sd}}{\Delta M}\right]e^{i\pi/4} \qquad (8)$$

where $(Re\lambda_t/s_1) = s_2(c_1s_2c_3 + c_2s_3c_\delta)$. At this point we note that the experimental bound[14] on the b-quark branching ratios into up versus charm quark final states $B(b \to u)/B(b \to c) < 0.09$ eliminates the sign ambiguity in $Re\lambda_t$ and hence in $s_2c_2s_3s_\delta$. Specifically, it was previously possible[11] that if $c_\delta < 0$ and $s_3 > s_2$, $Re\lambda_t$ could become sufficiently negative that the entire coefficient of $s_2c_2s_3s_\delta$ in Eq. (8) would become negative. Then $s_2c_2s_3s_\delta$ must also become negative to preserve the observed phase of ϵ. However, the constraint $B(b \to ue\nu)/B(b \to ce\nu) < 0.09$ requires[15] $|U_{ub}|^2 < 0.09 \times 0.41\ |U_{cb}|^2$ or that $s_3^2 < s_2^2 + 2s_2s_3c_\delta + s_3^2$, which in turn requires $s_2^2 + 2s_2s_3c_\delta > 0$. Thus, if $c_\delta < 0$ we observe that $Re\lambda_t/s_1 \simeq s_2^2 + s_2s_3c_\delta > s_2^2 + 2s_2s_3c_\delta > 0$ and therefore $Re\lambda_t$ is positive. ($Re\lambda_t$ is trivially positive if $c_\delta > 0$.) This fixes the sign of $s_2c_2s_3s_\delta$ *also to be positive.*[16] It now follows that the parameter ξ is negative. This comes from the known phase of C_6 ($ImC_6/ReC_6 < 0$) and the requirement that C_6Q_6 contribute *constructively* to A_0 which is positive by definition – i.e., that "penguins" contribute to a $\Delta I = \frac{1}{2}$ enhancement rather than a suppression. This in turn fixes ϵ'/ϵ through Eq. (5) to be *positive.*[16] Therefore we drop the modulus on $|\epsilon'/\epsilon|$ and treat ϵ'/ϵ as a positive quantity. Now, since ξ is negative, its presence in Eq. (8) only strengthens the bound on $s_2c_2s_2s_\delta$ which one obtains by dropping it. (It is in any case negligible in the domain we are consider of a lower bound on ϵ'/ϵ and hence on ξ.) Therefore we drop the term involving ξ, and use the experimental values for ϵ and s_1^2 to obtain

$$s_2c_2s_3s_\delta \geq \frac{2.4 \times 10^{-2}GeV^2}{\left[-0.7m_c^2 + 0.6m_t^2\left(\frac{Re\lambda_t}{s_1}\right) + 0.4m_c^2\ell n\,\frac{m_t^2}{m_c^2}\right]}\ . \qquad (9)$$

A lower bound on $s_2c_2s_3s_\delta$ obviously requires an upper bound on $m_t^2Re\lambda_t$, for which we turn to $K_L \to \mu\mu$ decay. The relevant constraint arises from requiring that the short-distance contribution to the $K_L \to \mu\mu$ branching ratio should not exceed the total minus the calculable absorptive contribution from $K_L \to \gamma\gamma \to \mu\mu$ with on-shell intermediate photons. The only uncertainty here is the long-distance contribution from the dispersive amplitude involving off-shell intermediate photons, which might interfere destructively with the short-distance contribution permitting the latter to be larger.[17] However, the present experimental situation[5] regarding the analogous decay $\eta \to \mu\mu$ (but without the short-distance contribution) indicates that the long-distance dispersive contribution is no bigger than the absorptive one (and consistent with zero). This strongly suggests that the long-distance dispersive contribution to $K_L \to \mu\mu$ should lead to at most a factor of two uncertainty in the short-distance contribution.

The lower bound on $s_2c_2s_3s_\delta$ which now follows from Eq. (9) and the upper bound on $m_t^2 Re\lambda_t$ from $K_L \to \mu\mu$ (including the full dependence on quark masses[12] and perturbative QCD[18]) is shown in Fig. 1. With maximum allowance for the uncertainties due to long-distance contributions to $K_L \to \mu\mu$, we observe that $s_2c_2s_3s_\delta \geq 2 \times 10^{-4}$. In constructing this bound we have used $m_c = 1.5$ GeV in the denominator of Eq. (9). However the value of m_c matters very little: using the upper bound on $m_t^2 Re\lambda_t$, the t quark contribution completely dominates CP violation in the $K^0 - \bar{K}^0$ mass matrix and the other terms in the denominator of Eq. (9).

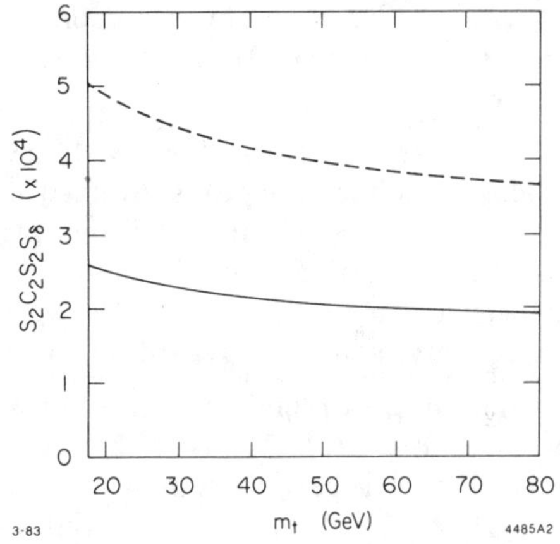

Fig. 1. Lower bound on the product $s_2c_2s_3s_\delta$ from $|\epsilon|$ and $K_L \to \mu^+\mu^-$ decay shown for $B = 0.33$: (a) with no allowance for long-distance dispersive contributions in $K_L \to \mu^+\mu^-$ (dashed); (b) with maximal allowance for long-distance dispersive contributions (solid). The bound scales inversely with B.

We are finally left to consider the coefficient function $Im\,\tilde{C}_6$ and the matrix element $< \pi\pi(I = 0)|Q_6|K^0 >$. There is very good agreement among the extant renormalization group analyses[6] for the imaginary part of C_6, which give $Im\,\tilde{C}_6 \approx -0.1$. This is because ImC_6 in particular gets generated at momentum scales between m_t and m_c and is therefore truly a short-distance effect susceptible to a leading logarithm calculation, and is quite stable with respect to variations in Λ_{QCD}. In particular, there is no strong dependence upon an infrared cutoff "μ" such as occurs for ReC_6, making the latter highly uncertain. It is because of this uncertainty in ReC_6 that we do not try to fit the matrix element $< \pi\pi(I = 0)|Q_6|K^0 >$ to the observed $\Delta I = 1/2$ rule, but use an independent bag model evaluation[19] which is smaller and hence more conservative. Combining this bag model matrix element with $Im\,\tilde{C}_6$ and our lower bound on $s_2c_2s_3s_\delta$, we obtain the lower bound shown in Fig. 2 from which we observe

Fig. 2. Lower bound on ϵ'/ϵ in the standard model as a function of m_t, shown for $B = 0.33$, $|Im\,\tilde{C}_6| = 0.1$ and $| < \pi\pi(I = 0)|Q_6|K^0 > | = 1.4 GeV^3$: (a) with no allowance for long-distance dispersive contributions in $K_L \to \mu^+\mu^-$ (dashed); (b) with maximal allowance for long-distance dispersive contributions (solid). The bound scales proportionally with $Im\,\tilde{C}_6$, $< 2\pi(I = 0)|Q_6|K^0 >$, and inversely with B.

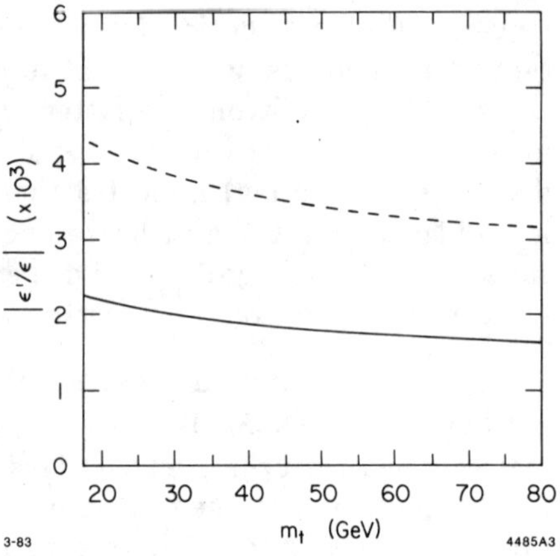

$$\frac{\epsilon'}{\epsilon} \geq 2 \times 10^{-3} \left(\frac{0.33}{B}\right) \left|\frac{Im\,\tilde{C}_6}{0.1}\right| \left|\frac{< \pi\pi(I = 0)|Q_6|K^0 >}{1.4 GeV^3}\right| \tag{10}$$

where the dependence of ϵ'/ϵ upon B, $|Im\,\tilde{C}_6|$, and the matrix element of Q_6 are explicitly shown. We conclude with a discussion of these uncertainties.

We have eliminated a major uncertainty by establishing a reliable lower bound on $s_2c_2s_3s_\delta$, the product characteristic of CP violation throughout the K-M model. This bound depends inversely on the parameter B, but if typical of other results which use $SU(3)$ and current algebra in their derivation, the value of 0.33 ought to be reliable to within $O(30\%)$. Moreover, our allowance for long-distance dispersive contributions to $K_L \to \mu\mu$ is probably an overestimate of the actual uncertainty, and so we regard our bound on $s_2c_2s_3s_\delta$ as conservative.

The largest remaining uncertainty concerns the $K \to \pi\pi$ matrix element of Q_6 in Eq. (10). We note that matrix elements of $(V - A) \times (V + A)$ type operators such as Q_6 are much more certain in the bag model than those of $(V - A) \times (V - A)$ operators: the integrals contributing enter with the same sign rather than opposite signs and there are no delicate cancellations which can lead to major uncertainties.[19] Moreover, we do not assume that the $\Delta I = 1/2$ rule is due to penguin contributions to the amplitude for $K \to \pi\pi$. This would require effectively "boosting" the matrix element of Q_6 (and hence ϵ'/ϵ) by at least a factor of two given most calculations[6] of ReC_6. We have taken a more conservative approach in seeking a lower bound on ϵ'/ϵ, by choosing an independent evaluation of the matrix element. While it is still possible that the actual matrix element is smaller than what we have used, smaller matrix elements make our understanding of the observed $\Delta I = 1/2$ dominance increasingly problematic.

REFERENCES

1. J. Christenson *et al.*, Phys. Rev. Lett. 13, 138 (1964).
2. T. C. Lee and C. S. Wu, Ann. Rev. Nucl. Sci. 16, 511 (1966).
3. M. Kobayashi and T. Maskawa, Prog. Theor. Phys. 49, 652 (1973).
4. F. J. Gilman and M. B. Wise, Phys. Lett. 83B, 83 (1979).
5. Particle Data Group, Phys. Lett. 111B (1982).
6. F. J. Gilman and M. B. Wise, Phys. Rev. D 20, 2392 (1979); B. Guberina and R. D. Peccei, Nucl. Phys. B163, 289 (1980); R.D.C. Miller and B.H.J. McKellar, Aust. J. Phys. 35, 235 (1982); F. J. Gilman and M. B. Wise, Phys. Rev. (in press), Ref. 9.
7. M. K. Gaillard and B. W. Lee, Phys. Rev. D 10, 897 (1974); J. Ellis, M. K. Gaillard and D. V. Nanopoulos, Nucl. Phys. B109, 213 (1976).
8. C. T. Hill, Phys. Rev. Lett. 97B, 275 (1980); J. S. Hagelin, Nucl. Phys. B193, 123 (1981).
9. F. J. Gilman and M. B. Wise, Phys. Lett. 93B, 129 (1980) and SLAC-PUB-2940 (1982), Phys. Rev. (in press).
10. These quantities are quite stable with respect to changes in quark masses and Λ_{QCD}, except for the dependence of η_1 on Λ_{QCD}. However this makes very little difference, as in the regime we are interested in the term proportional to m_c^2 is negligible compared to that proportional to m_t^2.
11. J. S. Hagelin, Phys. Rev. D 23, 119 (1981).
12. T. Inami and C. S. Lim, Prog. Theo Phys. 65, 297 (1981).
13. J. F. Donoghue *et al.*, Phys. Lett. 119B, 412 (1982).
14. B. Gittelman, CLEO Collaboration, and P. Franzini, CUSB Collaboration, papers given at the XXI International Conference on High Energy Physics, Paris, 26-31 July 1982.
15. M. K. Gaillard and L. Maiani, in Quarks and Leptons, Cargese 1979, eds. M. Levy, J. L. Basdevant, D. Speiser, J. Weyers, R. Gastmans, and M. Jacob (Plenum Press, New York, 1980) p. 433.
16. Since the θ_i are chosen to lie in the first quadrant by convention, this fixes s_δ to be positive. We emphasize that the positivity of ϵ'/ϵ distinguishes the K-M model from models where CP violation is predominantly in the $\Delta S = 1$ interaction [such as in the model of S. Weinberg, Phys. Rev. Lett. 37, 657 (1976)] where ϵ'/ϵ is negative, see J. S. Hagelin, Phys. Lett. 117B, 441 (1982).
17. V. Barger *et al.*, Phys. Rev. D 25, 1860 (1982).
18. J. Ellis and J. S. Hagelin, Nucl. Phys. B217, 189 (1983).
19. J. F. Donoghue *et al.*, Phys. Rev. D 23, 1213 (1981).

CALCULATING $|\varepsilon'/\varepsilon|$ IN $SU(2)_L \times SU(2)_R \times U(1)$ CAUCE MODELS

Gustavo C. Branco[†]
Virginia Polytechnic Institute and State University
Blacksburg, VA 24060

ABSTRACT

A calculation of the $|\varepsilon'/\varepsilon|$ ratio in electroweak models based on $SU(2)_L \times SU(2)_R \times U(1)$ is presented. We show that there is a class of theories where $|\varepsilon'/\varepsilon|$ is either vanishing or very small ($|\varepsilon'/\varepsilon| \approx 10^{-4}$). From the known strength of CP violation in the neutral kaon system, we obtain a constraint on the mass of the righthanded gauge boson: $M^2_{W_R} \approx 10^5 M^2_{W_L}$.

INTRODUCTION

At present there are various possible ways of introducing CP non-conservation [1] in electroweak gauge theories. It turns out that the ratio $|\varepsilon'/\varepsilon|$ can play a crucial role in distinguishing experimentally among the various theoretical schemes. In fact, it has been recently emphasized [2] that in the framework of $SU(2) \times U(1)$, the three scalar doublets model with natural flavor conservation and spontaneous CP violation [3] predicts values of $|\varepsilon'/\varepsilon|$ (i.e. $|\varepsilon'/\varepsilon| \gtrsim .02$) at the limit of inconsistency with present experimental data ($|\varepsilon'/\varepsilon| = .003 \pm .015$). The Kobayashi-Maskawa (K.M.) model [4] leads also to a non-vanishing value [5] for $|\varepsilon'/\varepsilon|$ ($.002 \leq |\varepsilon'/\varepsilon| \leq .02$). A substantially smaller value for $|\varepsilon'/\varepsilon|$ can be obtained in the framework of $SU(2)_L \times SU(2)_R \times U(1)$ gauge theories. These various predictions for $|\varepsilon'/\varepsilon|$ will be tested by experiments of improved accuracy which are underway or being planned [6].

The introduction of CP non-conservation in theories based on $SU(2)_L \times SU(2)_R \times U(1)$ was first done [7] in the context of four quarks. It was then shown that ε'/ε would automatically vanish in the case of four quarks, if one neglects W_L-W_R mixing and scalar boson interactions. It has been recently pointed out [8] that there is an algebraic mechanism that naturally leads to a vanishing $|\varepsilon'/\varepsilon|$ ratio, even when an arbitrary number of generations are considered. This result depends crucially on the assumption of spontaneous CP violation and manifest left-right symmetry. If the constraint of manifest left-right symmetry is not introduced, then

[†]Permanent address: Instituto Nacional de Investigacão Científica-Física Teórica e Métodos Matemáticos, 1699 Lisboa, Codex, Portugal.

$|\varepsilon'/\varepsilon|$ is non-vanishing, but very small ($|\varepsilon'/\varepsilon| \approx 10^{-4}$) due to an enhancement [9] of the $\Delta S = 2$ amplitude.

AN ALGEBRAIC MECHANISM

In this section, we will consider a gauge theory based on $SU(2)_L \times SU(2)_R \times U(1)$, with manifest left-right symmetry and spontaneous CP violation. It can be shown [8] that in this case the left-handed (K_L) and right-handed (K_R) mixing matrices entering in the charged weak currents are related by:

$$K_R = J^u K_L^* J^{d^\dagger} \tag{1}$$

where $J^{u,d}$ are, in general, unrelated unitary diagonal matrices:

$$J^u = \delta_{ij} \exp(i\chi_i) \ ; \quad J^d = \delta_{ij} \exp(-i\psi_i) \tag{2}$$

We will consider an arbitrary number of fermion generations but will assume that K_L can be made real by an appropriate choice of the phases of the fermion fields [10]. We will evaluate the effective $\Delta S = 1$ Hamiltonian, neglecting possible W_L-W_R mixing [11] as well as contributions arising from Higgs boson exchanges. The dominant

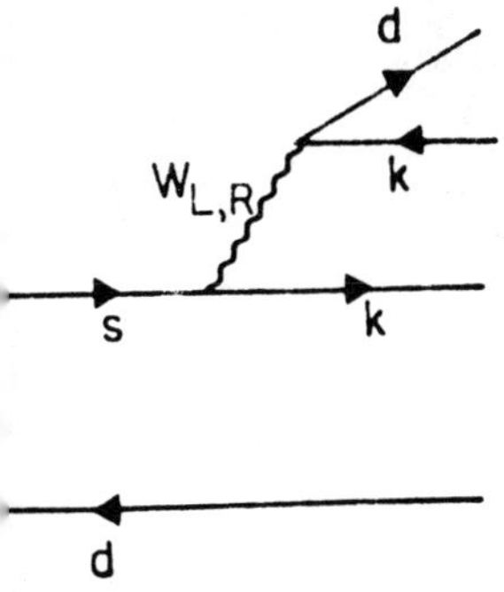

Figure 1

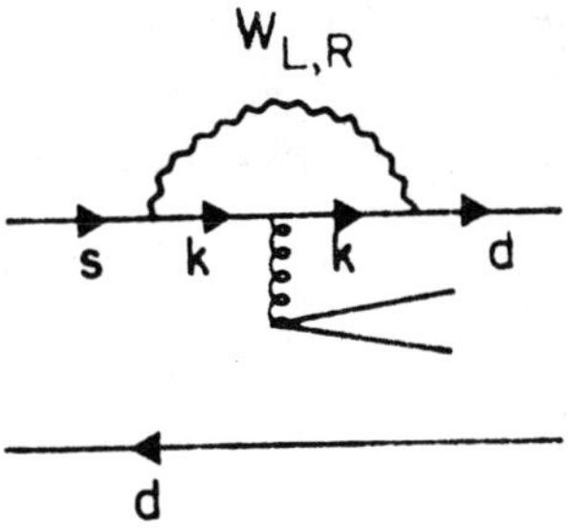

Figure 2

diagrams contributing to the $\Delta S = 1$ Hamiltonian are those of figs. 1 and 2. Using (1), (2), the contribution $G^{(i)}$ of graph (i) can be written as:

$$G^{(i)} = \sum_m k_{m2}\, k_{m1}\, [g_{Lm}^{(i)} + g_{Rm}^{(i)}\, e^{i(\psi_2-\psi_1)}] \tag{3}$$

where $k_{m\ell} \equiv (K_L)_{m\ell}$ are real numbers, while $g_{Lm}^{(i)}$, $g_{Rm}^{(i)}$ stand for the contributions from graph (i) with a Q = 2/3 quark (all mixing angles excluded). It is clear that contributions to the $K \to 2\pi$ transition due to W_L and W_R exchange are related by:

$$<2\pi|\, g_{Rm}^{(i)}\, |K> = - \frac{M_{W_L}^2}{M_{W_R}^2} <2\pi|\, g_{Lm}^{i}\, |K> \tag{4}$$

From (3), (4), it follows that:

$$A^{(i)} = A_L^{(i)} \left[1 - \frac{M_{W_L}^2}{M_{W_R}^2}\, e^{i(\psi_2-\psi_1)} \right] \tag{5}$$

where $A^{(i)}$ denotes the contribution of graph (i) to the $K \to 2\pi$ transition. From (5) one can immediately conclude that the $\Delta I = 1/2$ and $\Delta I = 3/2$ amplitudes have the same phase, thus implying $\varepsilon' = 0$.

AN ESTIMATE OF M_{W_R}

In order to estimate the mass of the right-handed gauge boson, we concentrate our attention now on the $K° - \bar{K}°$ system. The dominant contribution to the $\Delta S = 2$ transition comes from the usual box diagrams, where W_LW_L, W_LW_R or W_RW_R are exchanged. Since only external legs in the box diagrams give rise to phases, we can write:

$$A(K^o-\bar{K}^o) = \sum_{k\ell} [a_{LL}^{k\ell} + a_{LL}^{k\ell}\, e^{i(\psi_2-\psi_1)} + a_{RR}\, e^{2i(\psi_2-\psi_1)}] \tag{6}$$

where $a_{LL}^{k\ell}$, $a_{LR}^{k\ell}$, $a_{RR}^{k\ell}$ denote the absolute value of the amplitudes corresponding to W_LW_L, W_LW_R or W_RW_R exchange. The sum over k, ℓ correspond to the various pairs of Q = 2/3 quarks appearing in the box diagrams. On dimensional grounds, one obtains:

$$a_{RR}^{k\ell} = \left(\frac{M_{W_L}^2}{M_{W_R}^2} \right)^2 a_{LL}^{k\ell} \tag{7}$$

$$a_{LR}^{k\ell} = -2r^{k\ell} \frac{M_{W_L}^2}{M_{W_R}^2} a_{LL}^{k\ell} \tag{8}$$

where $r^{k\ell}$ are enhancement factors which have been recently calculated [9]. Some of these factors turn out to be surprisingly large [12]:

$$r^{cc} \geq 2 \times 10^2 \tag{9}$$

At this point, it is worthwile emphasizing that having $r^{k\ell} \neq 1$ is crucial in order to obtain CP non-conservation in the K°-$\bar{K}$° system. Indeed, if $r^{k\ell} = 1$ one would obtain:

$$A(K°-\bar{K}°) \approx \sum_{k\ell} a_{LL}^{k\ell} \left[1 - \frac{M_{W_L}^2}{M_{W_R}^2} e^{i(\psi_2-\psi_1)}\right]^2 \tag{10}$$

In that case, a suitable redefinition of the K°, $\bar{K}$° phases would make $<2\pi, I = 0, 2|H_{eff}|K°>$ and $A(K°\bar{K}°)$ simultaneously real, and consequently all CP violating effects would effectively vanish.

In order to estimate M_{W_R}, consider the usual parameter which measures the CP violation strength:

$$|\varepsilon| \cong \frac{1}{2\sqrt{2}} \frac{\text{Im } A(K°-\bar{K}°)}{\text{Re } A(K°-\bar{K}°)} \tag{11}$$

in the basis where $<2\pi, I = 0|H_{eff}^{\Delta S=1}|K°>$ is made real. From (3.9), (3.10) one obtains:

$$|\varepsilon| \approx \sin\delta \frac{a_{LR}}{a_{LL}} \approx \sin\delta \frac{M_{W_L}^2}{M_{W_R}^2} 2r \tag{12}$$

where δ denotes a typical phase associated with the right-handed mixing matrix and r is a typical enhancement factor introduced in eq. (8). If we assume $\delta = O(1)$, eq. (12) gives an estimate for M_{W_R}:

$$M_{W_R}^2 \approx 10^5 M_{W_L}^2 \tag{13}$$

CONCLUDING REMARKS

We have shown that there is a simple algebraic mechanism which makes $|\varepsilon'/\varepsilon|$ to vanish naturally in gauge theories based on $SU(2)_L \times SU(2)_R \times U(1)$, with manifest left-right symmetry and spontaneous CP violation. If these last constraints are not imposed, but if we keep the requirement that the breaking of CP invariance only arises from W_R interactions (i.e. K_L orthogonal), then the ratio $|\varepsilon'/\varepsilon|$ will no longer vanish, but a very small value is obtained:

$$|\varepsilon'/\varepsilon| \approx 10^{-4} \tag{14}$$

The result in eq. (14) is a rough estimate [8] and reflects the fact that there is an enhancement [9] in the $\Delta S = 1$ processes. As we have previously emphasized, we have neglected throughout our calculations, possible effects arising from W_L-W_R mixing [11]. An interesting alternative, where the left-right mixing can actually play a crucial role in generating CP breaking, has been recently suggested by Chang [12].

ACKNOWLEDGMENT

The work reported here was done in collaboration with J. M. Frère and J. M. Gérard. This work is supported in part by the US Department of Energy under Contract No. DE-AS05-80ER10713.

REFERENCES

1. For a review, see L. Wolfenstein, talk at this symposium.
2. A. I. Sanda, Phys. Rev. D23, 2647 (1981). N. G. Despande, Phys. Rev. D23, 2653 (1981). J. F. Donoghue, J. S. Hagelin and B. R. Holstein, Phys. Rev. D25, 195 (1982). D. Chang, Phys. Rev. D25, 1318 (1982). See however: Y. Dupont and T.N. Pham, CNRS report No. A498-0482 (1982).
3. G. C. Branco, Phys. Rev. D22, 2901 (1980).
4. M. Kobayashi and T. Maskawa, Prog. Theor. Phys. 49, 652 (1973).
5. F. Gilman and M. B. Wise, Phys. Lett. 83B, 83 (1979). B. Guberina and R. D. Peccei, Nucl. Phys. B1, 289 (1980). F. Gilman and J. Hagelin, SLAC-PUB-3087 (1983).
6. For a review, see B. Winstein, talk at this symposium.
7. R. N. Mohapatra and J. C. Pati. Phys. Rev. D11, 566 (1975).
8. G. C. Branco, J. M. Frère and J. M. Gérard, CERN preprint Ref. Th. 3406 (1982), to appear in Nuclear Physics B.
9. G. Beall, M. Bander and A. Soni, Phys. Rev. Lett. 48, 843 (1982).
10. This can be easily implemented in a technically natural way. See Ref. 8.
11. For an analysis of the calculability of W_L-W_R mixing, see G. C. Branco and J. M. Gérard, Phys. Lett. 124B, 415 (1983).
12. D. Chang, Nucl. Phys. B214, 435 (1983).

LEFT-RIGHT SYMMETRY AND CP-VIOLATION

Rabindra N. Mohapatra*
University of Maryland, College Park, MD 20742

and

Goran Senjanović
Brookhaven National Laboratory, Upton, N.Y. 11973

ABSTRACT

A review of left-right symmetry and its possible connection to CP-violation is presented. We also summarize the experimental situation regarding M_R, the mass scale of parity restoration and discuss the general features of quark mass matrices in these theories.

INTRODUCTION

Weak interactions, both in the realm of charged and neutral currents, manifestly violate parity -- this feature is a central property of the standard electro-weak models and is assumed apriori. The simplest way to understand the origin of parity violation leads naturally to the spontaneous breakdown of left-right symmetry, incorporated in the gauge theory based on $SU(2)_L \times SU(2) \times U(1)_{B-L}$ group.[1] This idea[1,2] is yet to be confirmed or rejected experimentally. The most important unanswered question in these theories is the value of the mass scale M_R at which parity is supposed to be restored, i.e. the mass of the right-handed gauge boson $W_R^{\pm}$. It is possible that CP-violation may play a crucial role in restricting M_R. Namely, there exists a particularly interesting possibility[3] that the smallness of CP violation in K decays is tied up to the maximality of parity violation through the connection of ε and ε' parameters to the ratio of $W_L^{\pm}$ and $W_R^{\pm}$ masses ($M_L \equiv M_{W_L}$, M_R).

In this short exposé, we review some general aspects of CP-violation in left-right symmetric theories, paying special attention to quark mass matrices and mixings. For the sake of completeness, in the next section, we first go over the existing experimental constraints on M_R and left-right mixing, in order that the reader know where we stand today and how these limits may be improved in the near future.

LEFT-RIGHT SYMMETRY: EXPERIMENTAL CONSTRAINTS

The basic features of left-right symmetry have been reviewed before; due to the lack of space we only summarize the experimental

*On leave from City College at New York, New York 10031

situation regarding both charged and neutral currents.

Charged Currents

The limits on M_R depend critically on the nature of neutrinos, which can be either Dirac or Majorana particles.

i) Dirac neutrinos: In this case, due to the left-right symmetry, $W_R^{\pm}$ participate in μ and β decays and hence, on the basis of the predominent V-A character of these decays, one can place a limit[4] $M_R \geq 3\ M_{W_L}$. We should point out, however, that in this case, the smallness of the neutrino mass is not easily understood, at least not naturally in the simple models.[5]

ii) Majorana neutrinos: This is a much more likely possibility, since in this case, the smallness of neutrino mass is connected to the maximality of parity violation in weak interactions,[6] through the relation $m_\nu \propto 1/M_R$. This suggests the right-handed neutrino (N_R) is very heavy: $M_{N_R} \simeq M_R$ in which case $W_R^{\pm}$ decouples from μ and β decays and low energy charged current processes put no limit on M_R at all![7]

Recently, in an interesting paper Beall et al.[8] suggested a possible limit on M_R coming from the smallness of the $K_L - K_S$ mass difference. They noticed that the contribution of $W_R^{\pm}$ to the $\Delta S = 2$ effective Hamiltonian, although down by $M_L{}^2/M_R{}^2$, comes with a large coefficient of order 400:

$$\frac{\langle H_{eff}(LR)\rangle}{\langle H_{eff}(LL)\rangle} = 8\log\left(\frac{m_c{}^2}{M_L{}^2}\right)\frac{\langle K^0|\bar{d}Ls\ \bar{d}Rs|K^0\rangle}{\langle K^0|(\bar{d}\gamma_\mu Ls)^2|\bar{K}^0\rangle}\left(\frac{M_L{}^2}{M_R{}^2}\right) \tag{1}$$

with $L,R \equiv \frac{1 \pm \gamma_5}{2}$, which with vacuum insertion becomes $\sim 430\ (M_L/M_R)^2$, leading to the limit: $M_R \geq 1.6$ TeV. This limit is independent of the nature of neutrinos, but it requires the same left and right mixing angles, which is expected in the simplest versions of the theory. A possible phenomenological loophole comes from the necessary tree-level Higgs contribution to the $\Delta S = 2$ effective Hamiltonian; for a light Higgs ($\leq$ 500 GeV), the limit on M_R could be as low as 200 GeV. The situation is summarized[9] in Figure 1, which gives the lower limit on M_R, for the fixed Higgs mass as a function of the t-quark mass. Low M_R solutions require partial cancellations of large Higgs and gauge boson contributions and so are somewhat unnatural; they are, however, phenomenologically consistent and must be kept in mind. They correspond to a narrow region for the quark mixing angles and could be independently tested.

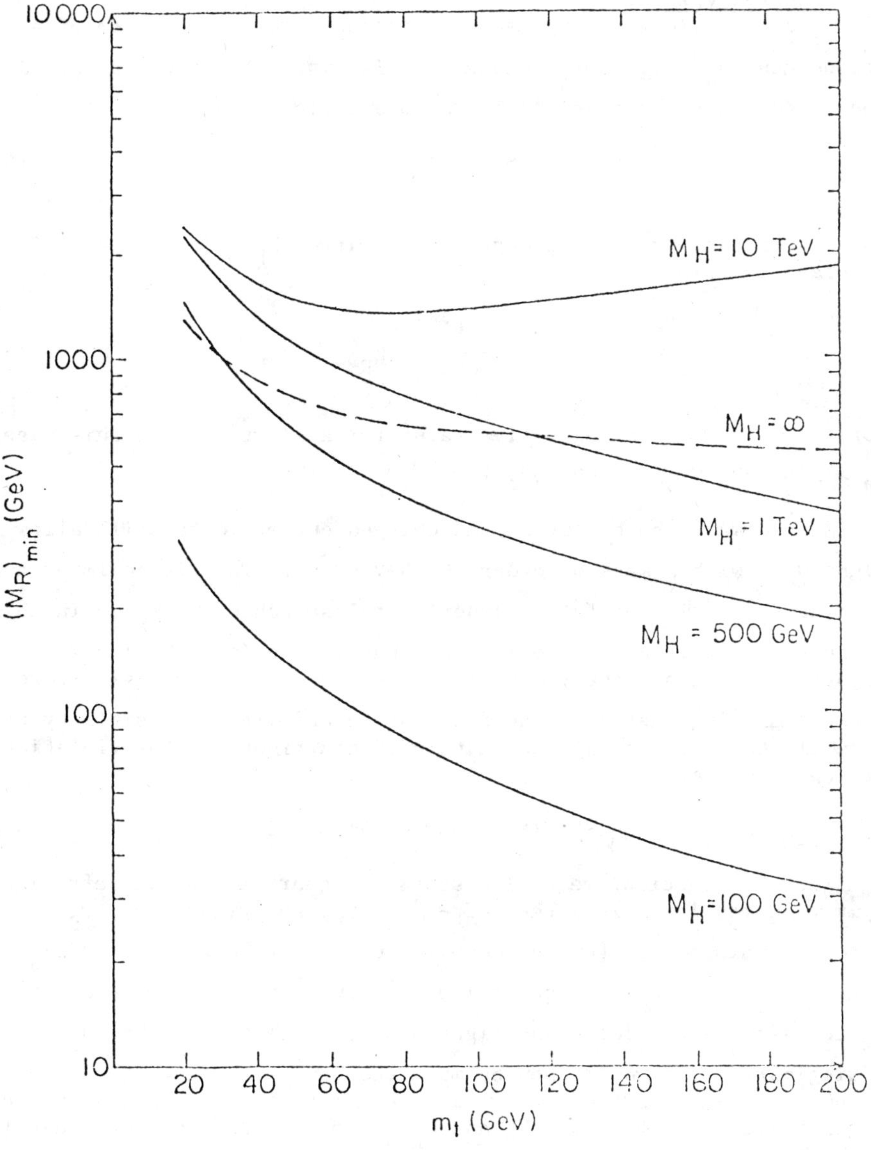

FIGURE 1

Lower limit on M_{W_R} from $K_L - K_S$ mass difference constraints for various values of the Higgs boson and top quark masses.

Neutral Currents

Indirect limits on M_R can be obtained from the presence of a second neutral gauge boson whose mass is typically of order M_R. A model independent analysis[10] leads to a limit

$$M_{Z_2} \gtrsim 200 \text{ GeV} \tag{2}$$

which then becomes a model dependent limit on M_R. In the case of Majorana neutrinos[6]

$$M_{Z_2}^{\,2} \simeq 2 \frac{\cos^2\theta_w}{\cos 2\theta_w} M_R^{\,2} \simeq 3 M_R^{\,2} \tag{3}$$

for $\sin^2\theta_w \simeq 0.25$, a suggestive value for a light W_R. In this case, $M_R \gtrsim 175$ GeV, $M_{Z_2} \gtrsim 3000$ GeV.

In summary, both neutral and charged current data still allow light $W_R^{\pm}$, with a mass of order 200 GeV or so. This is an important conclusion in view of future experimental searches for $W_R^{\pm}$. In this connection, it is useful to mention that in the (probable) case of Majorana neutrinos, the production of $W_R^{\pm}$ has very clear signatures with no missing energy in the final state and with a possibility of directly testing the Majorana nature of neutrinos and the violation of lepton number conservation.[12]

QUARK MIXING AND CP-VIOLATION

In this section, we will discuss the quark mixing in left-right symmetric models. Since the left- (Q_L, ψ_L) and right- (Q_R, ψ_R) handed fermions transform as doublets under the $SU(2)_L$ and $SU(2)_R$ weak groups, one will in general expect two mixing matrices U_L and U_R to describe the left- and right-handed currents and the number of mixing parameters would then be more than twice the number in the standard model. However, if we define parity transformation on the Higgs fields ϕ as $\phi \leftrightarrow \phi^+$, U_R and U_L become related to each other in one of the following ways[13]

(a) $\quad U_L = U_R$ (Manifest Left-Right Symmetry) (4a)

(b) $\quad U_L = U_R^*$ (Pseudo-Manifest L-R Symmetry) (4b)

If the Yukawa coupling matrices are complex and $\langle\phi\rangle$ is real, we get $M_{u,d} = M_{u,d}^{\dagger}$ which leads to case (a). On the other hand, if the

Yukawa coupling matrices are real and CP-violation is spontaneous i.e., $<\phi>$ = Complex, this implies $M = M^T$, which leads to case (b), as was shown in Ref. 13. Case (b) can also arise with real $<\phi>$, if under left-right transformation, we let $Q_L \to Q_R^c$ and $\phi \to \tilde{\phi}^+$ and the Yukawa couplings are chosen complex.

We wish to point out that, while the straightforward definition of the P-transformation on ϕ leads to constraints on U_R, it is possible to define left-right transformations that can leave U_R-completely free (see Ref. 13 for such a transformation).

In the rest of this note, we concern ourselves with case (b), which leads to an interesting new class of models for CP-violation;[3] in the next section, we discuss its implication for K-decays.

For case (a), suffice it to say that it is the left-right symmetric extension of the Kobayashi-Maskawa scheme, with small departures of order $M_L{}^2/M_R{}^2$. The main exciting feature of this type of model is that they may offer a natural solution of the strong CP problem.[14]

CP-VIOLATION AND K-DECAYS

To begin with, we remind the reader that in the standard $SU(2)_L \times U(1)$ model, the number of significant phases, $n_{cp} = \frac{(N-1)(N-2)}{2}$ where N is the number of fermion generations. So, one needs at least 3-generations to get a model of CP-violation. For the left-right symmetric models, on the other hand, the number of significant phases in both the left-(n_L) and the right-(n_R) handed sectors are given by:

$$n_L = \frac{(N-1)(N-2)}{2}$$

$$n_R = \frac{1}{2} N\ (N+1) \qquad (5)$$

So, we see that, even for two generations, there exist three significant phases (n_L=0 and n_R=3), which reduces to two if $W_L - W_R$ mixing is ignored. This feature was exploited by Mohapatra and Pati[3] to propose a model for CP-violation with two generations. Denoting $P \equiv (u,c)$ and $N \equiv (d,s)$, we can write the weak interaction Lagrangian for this model as:

$$\mathcal{L}_{wk} = \frac{g}{2\sqrt{2}} [\bar{P}_L\gamma_\mu U_L N_L W_L{}^{+\mu} + \bar{P}_R\gamma_\mu U_R N_R W_R{}^{+\mu} + h.c.] \qquad (6)$$

where

$$U_L = \begin{pmatrix} \cos\theta & \sin\theta \\ -\sin\theta & \cos\theta \end{pmatrix} \qquad (7)$$

$$U_R = e^{i\tau_3\alpha} U_L e^{-i\tau_3\alpha} e^{i\delta} \quad (8)$$

[τ_3 is the diagonal Pauli Matrix].

Ignoring Higgs- and W_L - W_R- mixing effects, we can write the $\Delta S = 1$ non-leptonic weak Hamiltonian in this model as:

$$H_w^{P.V.} = \frac{G_F}{2\sqrt{2}} \sin 2\theta\, 0_1^{(+)} (1 - \eta e^{i\delta}) + h.c. \quad (9a)$$

$$H_w^{P.C.} = \frac{G_F}{2\sqrt{2}} \sin 2\theta\, 0_2^{(+)} (1 + \eta e^{i\delta}) + h.c. \quad (9b)$$

where $\delta = -2\,\alpha_2$ and $0_i^{(+)}$ includes both the long distance and the Penguin contributions to the effective non-leptonic weak Hamiltonian. This form for the weak Hamiltonian leads to

$$\frac{M(K^0_2 \to \pi_i\pi_j)}{M(K^0_1 \to \pi_i\pi_j)} = + i\eta \sin\delta; \; (i,j = +- \text{ or } oo) \quad (10a)$$

and

$$\frac{M(K^0_1 \to \pi_i\pi_j\pi_k)}{M(K^0_2 \to \pi_i\pi_j\pi_k)} = -i\eta \sin\delta \quad (10b)$$

From these we obtain,

$$\eta_{+-} = e^{i\pi/4} \left[\frac{\text{Im } M_{K^0-\bar{K}^0}}{\text{Re } M_{K^0-\bar{K}^0}} + \eta \sin\delta \right] \quad (11a)$$

and

$$\eta_{+-} - \eta_{+-o} = 2\eta \sin\delta \quad (11b)$$

Thus, in the approximation that we ignore Higgs and W_L - W_R mixing effects, we obtain $\varepsilon' = 0$ and $\eta_{+-} - \eta_{+-o} = 2\eta \sin\delta$; the former coinciding with the superweak result, the latter potentially different from the superweak prediction. To discuss the second prediction, which could be tested by intense Kaon beams, we have to study $M_{K^0-\bar{K}^0}$.

In the left-right models, $M_{K^0-\bar{K}^0}$ receives contributions from the W_L-W_L, W_L-W_R[8] and Higgs[9] exchange graphs all of which are potentially large and can be written symbolically as

$$M_{K^0-\bar{K}^0} = f_{LL}\ [1 + \eta e^{i\delta} A_{LR} + e^{i\delta_2} A_H] \ldots \tag{12}$$

The absolute values of A_{LR} and A_H are much larger than one (e.g. in the vacuum saturation approximation, $A_{LR} \simeq -430$). The complexion of the CP-violating effect will depend on whether A_{LR} and A_H cancel each other in $M_{K^0-\bar{K}^0}$ (case a) or not case b).

Case (a): In this case, $\varepsilon \simeq \eta \sin\delta$ and $|\eta_{+-} - \eta_{+-o}| \simeq \eta_{+-}$, a prediction that can be tested by the next generation of Kaon-beam experiments.

Case (b): In this case, A_{LR} dominates η_{+-} and $A_{LR}\eta \gg \eta$; we obtain $\varepsilon \simeq \eta A_{LR} \sin\delta \approx \sin\delta$ and $|\eta_{+-} - \eta_{+-o}| \lesssim (10^{-2}$ to $10^{-3}) \times |\eta_{+-}|$ which is like the superweak prediction.

Note that in either case, CP and P-violation are related to each other, which is a unique feature of this model.

EXTENSION TO SIX QUARKS

Straightforward extensions of this idea to the case of more than two generations are obtained by noting the form of U_L and U_R in Eqs. 7 and 8. If we choose U_L = real and $U_R = K_1 U_L K_2$ where $K_{1,2}$ are diagonal unitary matrices, CP- and P-violation remain related to each other and also other predictions of the two generation model remain valid.[15]

Another way to extend the relation between CP- and P-violation for more than two generations has been discussed by D. Chang in these proceedings, where the key is to relate left-right mixing to the CP-violating phase in weak currents.

Finally, we wish to note that inclusion of W_L-W_R mixing ($\tan\zeta$) and Higgs exchange effects generates a non-vanishing ε' in this model, which is given by

$$\left|\frac{\varepsilon'}{\varepsilon}\right| \frac{\tan\zeta}{20} \lesssim 10^{-3} - 10^{-4} \quad \text{(case b)} \tag{13a}$$

$$\frac{\tan\zeta}{\eta}\,\frac{1}{20} \lesssim 10^{-3} \qquad \text{(case a)} \tag{13b}$$

This also leads to a prediction for the electric dipole moment of the neutron of $d_n^e \simeq 10^{-26}$ e-m.[16]

Another noteworthy feature of this kind of model is possible large CP-violation for semileptonic D meson decays[17] (for light W_R), a distinguishing feature from the Kobayashi-Maskawa and Higgs models.

One of us (R.N.M.) wishes to acknowledge the support of the National Science Foundation and the other (G.S.) acknowledges the support of the Department of Energy.

REFERENCES

1. J.C. Pati and A. Salam, Phys. Rev. D10, 275 (1974); R.N. Mohapatra and J.C. Pati, Phys. Rev. D11, 566; 2558 (1975). G. Senjanović and R.N. Mohapatra, Phys. Rev. D12, 1502 (1975).
2. For a review of left-right symmetry, see G. Senjanović, Nucl. Phys. B153, 334 (1979).
3. R.N. Mohapatra and J.C. Pati, Phys. Rev. D11, 566 (1975).
4. M.A.B. Bég, R.V. Budny, R.N. Mohapatra and A. Sirlin, Phys. Rev. Lett. 38, 1252 (1977).
5. See G.C. Branco and G. Senjanović, Phys. Rev. D18, 1621 (1978).
6. R.N. Mohapatra and G. Senjanović, Phys. Rev. Lett. 44, 912 (1980); and Phys. Rev. D23, 165 (1981).
7. For a situation in which M_{N_R} is substantially lighter, see M. Gronau and S. Nussinov, FERMILAB-PUB-82/52-THY (1982).
8. C. Beall, M. Bander and A. Soni, Phys. Rev. Lett. 48, 848 (1982).
9. R.N. Mohapatra, G. Senjanović and M. Tran, Phys. Rev. D (to appear).
10. V. Barger, K. Whisnant, N. Deshpande and R. Johnson, Wisconsin preprint (1983).
11. T. Rizzo and G. Senjanović, Phys. Rev. D24, 704 (1981).
12. W.Y. Keung and G. Senjanović, BNL preprint (1983).
13. R.N. Mohapatra, F.E. Paige and D.P. Sidhu, Phys. Rev. D17, 2642 (1978); and R.N. Mohapatra, "New Frontiers in High Energy Physics", ed. A. Perlmutter and L. Scott (Plenum Press, New York, 1979).
14. M.A.B. Bég and H.S. Tsao, Phys. Rev. Lett. 41, 278 (1978); R.N. Mohapatra and G. Senjanovic, Phys. Lett. 79B, 283 (1978).
15. G. Branco, these proceedings.
16. G. Beall and A. Soni, Phys. Rev. Lett. 47, (1981).
17. G. Kane and G. Senjanović, Phys. Rev. D25, 173 (1982).

CP-VIOLATION THROUGH HIGGS EXCHANGE

Nilendra G. Deshpande
Institute of Theoretical Science and Department of Physics
University of Oregon, Eugene, Oregon 97403

ABSTRACT

We review the Weinberg model of CP nonconservation through the exchange of Higgs bosons. When possible long distance contribution to $K-\bar{K}$ mass matrix are included, the model is barely compatible with the limit for ε'/ε obtained from experiment. We also discuss the electric dipole moment (e.d.m.) of the neutron in this model.

I. INTRODUCTORY REMARKS

As first discussed by Kobayashi and Maskawa[1], CP-violation can occur in gauge theories in three ways: (a) mixing angles with a complex phase in six (or more) quark model (b) additional gauge bosons as in Left-Right symmetric theories (c) through more Higgs doublets. Here we discuss the last possibility. There are two ways in which the model can be implemented. These we call Type A and Type B models.[2] The type B has natural flavor conservation in the neutral sector, and is more attractive theoretically.

II. TYPE A MODELS

These models require at least two Higgs doublets with arbitrary couplings to fermions. The neutral components of the Higgs doublets have flavor changing couplings, and consequently their masses have to be very heavy. Models of this type were suggested by T. D. Lee[3] and developed by P. Sikivie[4] and Lahanas and Vayonakis[5]. We do not consider models of this type in detail here but merely summarize their general features.

1. CP can be broken spontaneously or intrinsically. CP-violations occur in both the Higgs sector and through mixing angles in the gauge sector.
2. CP-violating Higgs interactions are mediated by neutral bosons. There is a direct $\bar{s}d \rightarrow (H_1,H_2) \rightarrow \bar{d}s$ vertex which has a CP violating phase. The masses of H_1 and H_2 lie in the TeV range barring accidental cancellation.
3. The contribution to ε'/ε through Higgs exchange is vanishingly small.
4. The e.d.m. of the neutron has an important contribution through the e.d.m. of u quark. The diagram involving $u \rightarrow t + H_{1,2} \rightarrow u$ should dominate. The contribution is expected to be of the order of $D_N \approx 10^{-26}$ e-cm.
5. Rare processes like $K_L \rightarrow \mu e$ and $\mu + X \rightarrow e + X$ are allowed at a B. R. of 10^{-13} or so.

The model has two serious weakness:

a) Very high masses of Higgs bosons which are barely compatible with Unitarity,

b) lack of definite Higgs-fermion couplings. The last feature can be corrected by imposing some discrete symmetries which might also explain the generation puzzle. One such an attempt is in progress by us.[6]

III. CLASS B MODEL

This type of model, which has flavor conservation in the neutral sector at the tree level, was first proposed by Weinberg.[7] The model needs at least three Higgs doublets. Discrete symmetries are imposed so that all up-type quarks get their masses from one Higgs doublet, while down-type quarks get their masses from a second doublet. CP can be broken spontaneously or intrinsically. If the breaking is spontaneous, the mixing angles in quark-gauge coupling are real.[8] This latter version is the model we will study because the CP nonconservation in the Higgs sector is then isolated from the CP nonconservation through the KM phase.

The Higgs interaction in this model is given by

$$L = \frac{\phi_1^-}{\lambda_1^*} \bar{d}_{iR} M^d (K^+)_{ij} u_{jL} + h.c. - \frac{\phi_2^+}{\lambda_2} \bar{u}_{iR} M_i^u K_{ij} d_{jL} + h.c. + \text{neutral currents} + \text{leptonic currents} \qquad (1)$$

here $u_i \equiv (u, c, t \ldots)$, $d_i = (d,s,b \ldots)$, M^u and M^d are diagonal quark matrices for up-type and down-type quarks, and K is the K-M matrix with real elements. The CP-violation arises due to the complex propagator

$$A = \frac{\langle 0|\phi_1^- \phi_2^+|0\rangle}{\lambda_1^* \lambda_2} \qquad (2)$$

IV. CP PHENOMENOLOGY IN THE K SYSTEM

We define K_L and K_S in standard manner as

$$|K\rangle_{S,L} = \left[\frac{1}{2(1+|\varepsilon|^2)}\right]^{\frac{1}{2}} \left[(1+\varepsilon)|K^\circ\rangle \pm (1-\varepsilon)|\bar{K}^\circ\rangle\right] \qquad (3)$$

Further, from CPT

$$\langle 2\pi, I|H_W|K^\circ\rangle = A_I e^{i\delta_I} \qquad (4)$$

$$<2\pi,\ I|H_w|\ \bar{K}^\circ> = A_I^* e^{i\delta_I} \tag{5}$$

for I = 0, 2. In the Wu-Yang phase convention $A_0 = A_0^*$. In this phase convention

$$\varepsilon \cong \frac{e^{i\pi/4}}{2\sqrt{2}} \varepsilon_m \text{ where } \varepsilon_m = \frac{\text{Im}M_{12}}{\text{Re}M_{12}} \tag{6}$$

and M_{12} is K, $\bar{K}$ mass matrix. We also have for ε'

$$\varepsilon' = \frac{i}{\sqrt{2}} e^{i(\delta_2-\delta_0)} \frac{\text{Im}A_2}{A_0} \tag{7}$$

and

$$\eta_{+-} = \frac{<\pi^+\pi^-|H_w|K_L>}{<\pi^+\pi^-|H_w|K_S>} \cong \varepsilon+\varepsilon' \tag{8}$$

$$\eta_{oo} = \frac{<\pi^o\pi^o|H_w|K_L>}{<\pi^o\pi^o|H_w|K_S>} \cong \varepsilon-2\varepsilon' \tag{9}$$

In many theoretical calculations we are not in the Wu-Yang convention. We can however, always make a phase rotation to reach this convention.

A standard convention in SU(3) will yield the matrix element with H_w^{CP+} to be real and H_w^{CP-} to be pure imaginary. In such a convention we have

$$\frac{\text{Im}<(2\pi,\ I = 0|H_w|K^\circ>}{\text{Re}<(2\pi,\ I = 0|H_w|K^\circ>} = \xi << 1 \tag{10}$$

Making a rotation $|K^\circ> \rightarrow e^{-i\xi} |K^\circ>$ we have[9]

$$\varepsilon = \frac{e^{i\pi/4}}{2\sqrt{2}} (\varepsilon_m + 2\xi) \tag{11}$$

$$\varepsilon' = \frac{i}{\sqrt{2}} e^{i(\xi_2-\xi_0)} \frac{I_m [e^{-i\xi} A_2]}{A_o} \tag{12}$$

Even in models where the CP nonconserving operator is predominately ΔI = 1/2 as in Weinberg model and $\text{Im}A_2 = 0$, we have

$$\frac{\varepsilon'}{\varepsilon} = -\frac{1}{20} \left(\frac{2\xi}{\varepsilon_m + 2\xi}\right) \tag{13}$$

V. PENGUIN CONTRIBUTION IN WEINBERG MODEL

The CP violating penguin contribution to $K \to 2\pi$ comes out to be very large in this model.[10] The $s \to d + g$ CP violating interaction is

$$L = i\, f\, \bar{d}_R \sigma^{\mu\nu} \frac{\lambda^a}{2} s_L F^a_{\mu\nu} \tag{14}$$

with

$$f = \frac{g_s}{32\pi^2} M_c^{\,2} M_s \cos\theta_c \sin\theta_c \,(\mathrm{Im}A)\left[\ell n \left(\frac{M_H^{\,2}}{M_c^{\,2}}\right) - \frac{3}{2}\right] \tag{15}$$

where we have assumed $m_{H_2}^2 \gg m_{H_1}^2 = m_H^2$.

Similarly, the $\Delta S = 2$ operator due to W-Higgs exchange box diagrams is

$$L = g\, \bar{d}_i \gamma_\mu s_{Rj}\, i\partial_\nu \bar{d}_j \gamma^\mu \gamma^\nu s_{Li} \tag{16}$$

with

$$g = \frac{G_F}{\sqrt{2}} \frac{1}{16\pi^2} M_c^{\,2} M_s \,(\cos\theta_c \sin\theta_c)^2\, (\mathrm{Im}A) \tag{17}$$

Using current algebra and vacuum saturation one now finds[10]

$$\varepsilon_m \cong \frac{3}{16\sqrt{2}} \left[\frac{M_K^{\,2}(\mathrm{Im}A)}{G_F}\right] \tag{18}$$

and

$$\xi \cong \left[\frac{M_K^{\,2}(\mathrm{Im}A)}{G_F}\right] \tag{19}$$

Thus

$$\varepsilon_m \ll \xi \tag{20}$$

and

$$\frac{\varepsilon'}{\varepsilon} \cong -\frac{1}{20} \tag{21}$$

This result has been greatly strengthened by Donoghue, Hagelin and Holstein[11] who have used the MIT bag model to evaluate matrix elements, and included t quark effects. They find $\xi > 7\varepsilon_m$ in all cases except one, where accidental cancellation of large terms might lower ξ. If this result for ε'/ε stands up, the model is already ruled out because of the experimental limit

$$\varepsilon'/\varepsilon = -.003 \pm .015 \tag{22}$$

We shall consider effects which might lower the estimate of ε'/ε in the next section.

VI. EFFECT OF LONG DISTANCE CONTRIBUTIONS

The estimates for ε'/ε above depended on the contribution to the mass matrix M_{12} arising solely through the short distance box diagram. Such a diagram neglects effects due to low lying resonant states, like π, η, ρ, etc. These states can contribute to both the real and the imaginary parts of M_{12}. We shall refer to these contributions as long distance contributions. It is known from earlier work[12] based on current algebra that the states π, η can make a substantial contribution to M_{12}. We can include these effects quite easily, following Hill[13], Wolfenstein[14], Chang[15], and Hagelin[16]. We divide M_{12} into short and long distance parts

$$M_{12} = M_{12}^{S} + M_{12}^{L} \tag{23}$$

We relate $\mathrm{Im} M_{12}^{L}$ to $\mathrm{Re}\, M_{12}^{L}$ by assuming each matrix element is related as in Eq. (10).

$$\mathrm{Im} \langle \mathrm{resonance}\, |H_w|\, K^\circ \rangle = \xi \mathrm{Re} \langle \mathrm{resonance}\, |H_w|\, K^\circ \rangle \tag{24}$$

Then

$$\mathrm{Im}\, M_{12}^{L} = -\, 2\xi\, \mathrm{Re}\, M_{12}^{L} \tag{25}$$

We further assume

$$\mathrm{Re}\, M_{12}^{L} = -z\, \mathrm{Re}\, M_{12} \tag{26}$$

where z is a phenomenological parameter. Then

$$(1+z)\mathrm{Re}\, M_{12} = \mathrm{Re}\, M_{12}^{S} \tag{27}$$

We shall define

$$\varepsilon_m^S = \text{Im } M_{12}^S \ / \ \text{Re } M_{12}^S \tag{28}$$

$$\varepsilon_m = \text{Im } M_{12} \ / \ \text{Re } M_{12} \tag{29}$$

Then it follows that

$$\varepsilon_m + 2\xi = (\varepsilon_m^S + 2\xi)(1+z) \tag{30}$$

$$\frac{\varepsilon'}{\varepsilon} = -\frac{1}{20} \frac{1}{(1+z)} \left[\frac{2\xi}{\varepsilon_m^S + 2\xi}\right] \tag{31}$$

A reasonable estimate for z would be[12]

$$-2 < z < 2$$

only z near 2 is allowed experimentally, and we have

$$\frac{\varepsilon'}{\varepsilon} < -.016 \tag{32}$$

Thus the model is barely allowed, and makes a prediction that ε'/ε is negative.

VII. NEUTRON ELECTRIC DIPOLE MOMENT

An expression for electric dipole moment of quark of type i is approximately

$$D_i = [\text{Im } A] \frac{1}{16\pi^2} M_i M_j^2 K_{ij}^2 e_j \left(\ell n \frac{M_H^2}{M_j^2} - \frac{3}{4}\right) \tag{33}$$

Since the long distance effects are important, one can estimate ImA from the value of ξ which is related to ε. Such a calculation depends to some degree on mass of t quark and mixing angles. A new calculation by Beall and me[17] shows, for a fairly general situation, the e.d.m. of N, D_N is

$$D_N > 3 \times 10^{-26} \text{ e-cm}$$

The present limit is $|D_N| < 4 \times 10^{-25}$ e-cm.

VIII. CONCLUSIONS

The Higgs model of CP violation is barely consistent with data when possible long distance effects are included. A slight improvement in the experimental situation in ϵ'/ϵ should be able to rule the model out. We have also looked at CP violating effects of the model in $K\ell_3$ decays, and find them to be too small to be seen experimentally.

REFERENCES

1. M. Kobayashi and T. Maskawa, Progr. Theoret. Physics. (Kyoto) 49, 652 (1973).
2. S. Pakvasa, AIP Conference Proceedings 68, 1165 (1981).
3. T. D. Lee, Phys. Rev. D8, 1226 (1973).
4. P. Sikivie, Phys. Lett. 65B, 141 (1976).
5. A. Lahanas and C. Vayonakis, Phys. Rev. D19, 2158 (1979); also J. Jacquot, Lett. Nuovo Cimento 26, 155 (1979).
6. T. Brown, N. G. Deshpande, and S. Pakvasa, to be published.
7. S. Weinberg, Phys. Rev. Lett. 37, 657 (1976).
8. G. C. Branco, Phys. Rev. Lett. 44, 504 (1980).
9. F. G. Gilman and M. B. Wise, Phys. Rev. D20, 2392 (1979).
10. N. G. Deshpande, Phys. Rev. D23, 2654 (1981) and A. I. Sanda, Phys. Rev. D23, 2647 (1981).
11. J. F. Donoghue, J. S. Hagelin, and B. R. Holstein, Phys. Rev. D25, 195 (1982).
12. P. Auvil and N. G. Deshpande, Nuovo Cimento 46, 766 (1966) and C. Itzksin, M. Jacob, and G. Mahoux, Nuovo Cimento Supp. 5, 978 (1967).
13. C. T. Hill, Phys. Lett. 97B, 275 (1980).
14. L. Wolfenstein, Nuclear Physics B160, 501 (1979).
15. D. Chang, Phys. Rev. D25, 1381 (1982).
16. J. S. Hagelin, Phys. Lett. 117B, 441 (1982).
17. G. Beall and N. G. Deshpande, to be published.

ON THE STRONG CP PROBLEM IN THEORIES WITH SEVERAL HIGGS DOUBLETS AND IN SUPERSYMMETRIC MODELS[1]

I. I. Bigi
Department of Physics
University of California, Los Angeles, CA 90024

and

Institut für Theoret. Physik E
RWTH Aachen, D5100 Aachen, F.R.G.

DEDICATED TO THE MEMORY OF J.J. SAKURAI

The strong CP problem which surfaced a few years ago[2] still awaits a clearly established solution. It can be stated as follows: attempting to include non-perturbative effects one uses an effective strong Lagrangian

$$\mathcal{L}_{eff} = \mathcal{L}_{QCD} + \frac{\theta}{32\pi^2} F\tilde{F} .$$

The experimental upper bound on the electric dipole moment of the neutron can be translated[3] into $\theta \lesssim 10^{-9}$. In QCD there is no apparent reason why this parameter should be so tiny. Even if one puts θ to zero in $\mathcal{L}_{eff}$ it will be renormalized by quantum corrections due to weak interactions. Actually there will be finite contributions to θ renormalization as well as infinite ones since CP violation is usually produced by hard, i.e. [mass] dimension four operators. It has been shown by Ellis and Gaillard[4] that in the Kobayashi-Maskawa ansatz of CP violation $\delta\theta$ receives its first finite contribution at the two-loop level: $\delta\theta \sim 10^{-16}$; infinities do not arise before the seven-loop level and it can be argued that a more complete theory will introduce a cut-off which renders such high order contributions harmless. Ellis and Gaillard have also given a rough estimate of $\delta\theta$ in the Weinberg model of CP violation[5] where they conclude that finite contributions to $\delta\theta$ arise on the one-loop level and are much too large; infinities occur already on the two-loop level. We performed a more detailed computation of $\delta\theta$ in the Weinberg ansatz for two reasons:

i) Although the result which we obtain is numerically very close to the rough estimate of Ellis and Gaillard, this is due to a cancellation of large factors.

ii) The calculation presented here will be an important ingredient in our later discussion on the impact of supersymmetry.

We find

$$\delta\theta(\text{1-loop}) = \frac{\sqrt{2}\,G_F}{8\pi^2}\,\text{Im}\,\alpha_1\beta_1^* \times \left\{ K_{\alpha\beta}M_\beta^D M_\beta^D K^+_{\beta\alpha} \left[\frac{M_1^2}{M_1^2-(M_\beta^D)^2}\log\frac{M_1}{M_\beta^D} - \frac{M_2^2}{M_2^2-(M_\beta^D)^2}\log\frac{M_2}{M_\beta^D} \right] + K^+_{\alpha\beta}M_\beta^U M_\beta^U K_{\beta\alpha}[U \leftrightarrow D] \right\}$$

where $M_1\,[M_2]$ denotes the mass of the charged Higgs field belonging to the SU(2) doublet $H_1\,[H_2]$ and $M^D\,[M^U]$ stands for the mass of the down [up] quarks d,s,b,...[u,c,t,...]; K is the Kobayashi-Maskawa (=KM) quark mixing matrix. Im $\alpha_1\beta_1^*$ describing mixing between H_1 and H_2 can be calibrated by attributing CP violation in K^0-$\bar{K}^0$ mixing solely to this source.[6] Inserting numbers we find

$$|\delta\theta(\text{1-loop})| \gtrsim 10^{-3}$$

from m (top) ~ 20-40 GeV. Such θ renormalization is obviously too large and, as already mentioned, it becomes even uncontrollable in two loops.

Supersymmetry (= SuSy) is not introduced to accommodate CP violation. However, with its more complex structure it allows for more sources of CP violation, even in the minimal SuSy extension of the Standard Model:[7] the squarks will have a "KM" matrix that can in general contain complex phases like their SuSy partners, the quarks. In addition the left-handed and right-handed squarks can mix leading to CP violation as in the Weinberg ansatz. Finally the higgsinos and gauginos can also mix thus exhibiting complex phases.[8]

SuSy per se does not solve the strong CP problem. However, putting θ to zero at the tree level -- perhaps by appealing to an embedding of SuSy into supergravity[9] -- makes better sense: with SuSy unbroken not even a finite renormalization of θ occurs, and with SuSy broken softly or spontaneously, only a finite renormalization will arise. Thus a perturbative expansion of δθ should offer a good semi-quantitative guideline.

The algebra of the one-loop calculation is quite analogous, *cum grano salis*, to the one mentioned above for the Weinberg ansatz. Assuming for example M(gaugino) ~ 1 TeV, M(higgsino), m(squark), Δm(gaugino) ~ 100 GeV and Δm(squark) ~ 10 GeV we find

$$\delta\theta(\text{1-loop}) \sim (10^{-6} - 10^{-4}) \times \text{ratios of mixing angles.}$$

These numbers are still much too large unless one postulates some rather extreme values for the plethora of mixing angles involved. From this discrepancy we conclude that (at least) one of the following alternatives should hold:

i) The strong CP problem is an artifact of our attempts at including non-perturbative effects.

ii) The relevant SuSy scale is not ~ 1 TeV, but much higher while at the same time gauginos, higgsinos and squarks are almost degenerate in mass on that scale:

$$\frac{\Delta M(\text{ino,sq})}{M(\text{ino,sq})} \ll 1.$$

Since $\delta\theta$ depends on mixing between "inos" or squarks it would be suppressed accordingly.

iii) Even with SuSy broken, some hidden symmetry is preserved to keep some relevant mixing angles or phases very small (or even zero).

iv) SuSy contains all the ingredients of having a Peccei-Quinn symmetry;[10,11] however, in general it is softly broken and therefore no axions are introduced.[9] But perhaps the Peccei-Quinn symmetry is only spontaneously broken making θ renormalization as discussed here irrelevant.

v) In some classes of spontaneous broken SuSy theories, one finds that the one loop contribution to $\delta\theta$ vanishes. Ellis, et al.[12] find that in these models, $\delta\theta \sim O(10^{-16})$ due to two-loop effects.

For a discussion of direct SuSy contributions to the electric dipole moment of the neutron, see M.B. Gavela's contribution to these Proceedings.

ACKNOWLEDGEMENTS

I gratefully acknowledge illuminating discussions with M.B. Gavela, H.E. Haber, R. Mohapatra and G. Senjanovic. It was a delight to participate in this Workshop and I am grateful to the organizers, in particular T. Goldman and H.E. Haber, for creating such an inspiring atmosphere.

REFERENCES

1. Work done in collaboration with R. Akhoury, UCLA preprint.
2. G. 't Hooft, Phys. Rev. Lett. 37, 8(1976); Phys. Rev. D14, 3432 (1976).
3. V. Baluni, Phys. Rev. D19, 2227 (1979).
4. J. Ellis and Mary K. Gaillard, Nucl. Phys. B150, 141 (1979).
5. S. Weinberg, Phys. Rev. Lett. 37, 657 (1976).
6. N.G. Deshpande, Oregon preprint OITS 196 (1982).
7. See e.g.: S. Weinberg, Phys. Rev. D26, 287 (1982).
8. See for similar ideas: F. del Aguila et al., HUTP-83/A017 (1983).
9. See e.g.: R.N. Mohapatra et al., BNL 32582 (1983).
10. R. Peccei and H. Quinn, Phys. Rev. Lett. 38, 1440 (1977); Phys. Rev. D16, 1791 (1977).

11. M. Dine et al., Nucl. Phys. B189, 575 (1981);
H.P. Nilles and S. Raby, Nucl. Phys. B198, 102 (1982).
12. John Ellis, Sergio Ferrara and D.V. Nanopoulos, Phys. Lett. 114B, 231 (1982).

CP-VIOLATION IN SUPERSYMMETRIC THEORIES

M.B. Gavela
Brandeis University, Waltham, MA 02254

ABSTRACT

We discuss sources of CP violation specific to supersymmetric theories. CP-violating phases may appear: a) in the mass matrices of the gauge and Higgs fermions; b) in the mixing between the scalar quarks; c) in the gauge couplings (we revisit the supersymmetric version of the Kobasyashi-Maskawa model). We estimate its effect on the electric dipole moment of the electron and the neutron. In some cases, it is found to be close to the experimental upper bounds.

I will report here on work done in several collaborations with F. del Águila, J.M. Frère, Howard Georgi, J.A. Grifols and A. Méndez.

We wish to begin by drawing your attention to a specific source of CP-violation[1] present in a large class of supersymmetric (SUSY) models, where supersymmetry is realized either globally or locally: In SUSY models we have at least four new neutral Majorana fermions, $\tilde{W}_3$, $\tilde{B}$, $\tilde{H}^0$, $\tilde{H}^{0\prime}$. They are the partners of the neutral gauge bosons (W_3, B) of $SU(2)_L \times U(1)$ and the neutral Higgs bosons $(H^0, H^{0\prime})$ of the two doublets H and H', which are required to give masses to the fermions.[2] These new fermions will mix, in principle, with non-absorbable phases inducing possible CP-violation effects. Similarly, CP-violating phases appear in the mixing between the super-symmetric partners of the charged $SU(2)_L$ gauge bosons and Higgs, $\tilde{H}^{\pm}$, $\tilde{W}^{\pm}$. The mass matrix for the neutral Majorana fields has the form

$$(\tilde{W}_3, \tilde{B}, \tilde{H}^0, \tilde{H}^{0\prime}) \left(\begin{array}{cc|cc} \tilde{M} & 0 & \multicolumn{2}{c}{o(M_W)} \\ 0 & \tilde{M}' & & \\ \hline & & 0 & m \\ \multicolumn{2}{c|}{o(M_W)} & m & 0 \end{array}\right) \begin{pmatrix} \tilde{W}_3 \\ \tilde{B} \\ \tilde{H}^0 \\ \tilde{H}^{0\prime} \end{pmatrix}$$

where m is arbitrary.

These mixings may induce sizeable contributions for the electric dipole moments (e.d.m.) of hadrons and leptons, at the one loop level. Natural values for the masses are

$$m_{sf_L}^2 \sim m_{sf_R}^2 << \tilde{M}^2 \sim \tilde{M}'^2 \sim (1\ \text{TeV})^2$$

where sf refers to the scalar fermion partners of the known fermions.

The resulting e.d.m.'s are

$$d_f \sim \frac{m_f(\mathrm{MeV})}{\tilde{M}^2(\mathrm{TeV})} \cdot \sin\delta \cdot 10^{-25} \text{ e-cm.}$$

$$d_{neutron} \simeq 10^{-24} \sin\delta \text{ e-cm.}$$

$$d_{electron} \simeq 10^{-25} \sin\delta \text{ e-cm.}$$

where sinδ parameterizes the amount of CP-violation present. Similar order of magnitude estimates should be expected from the exchange of the charged gauge and Higgs fermions.

A second source of CP-violation in some models with soft SUSY breaking has been analyzed by W. Buchmüller and D. Wyler.[3] (This kind of breaking may be present in N=1 supergravity models.[4]) In the case of a non-vanishing gluino mass, the e.d.m. of the neutron is caused by the mixing of right- and left-handed scalar quarks.[3,5] One loop gluino exchange then induces:

$$d_{neutron} \lesssim 4 \times 10^{-25} \text{ e-cm.}$$

We find[6], in an analogous way,

$$d_{electron} \lesssim 10^{-24} \text{ e-cm.}$$

In fact, from the present experimental bounds, these calculations provide interesting bounds on products of CP-violating phases times mass differences in the scalar quark and lepton sector.

A third source of CP-violation appears in the gauge couplings of the fermions and scalar fermions to the gauge bosons and fermions. The effect is just the supersymmetric extension of the Kobayashi-Maskawa model and has been previously analyzed in detail by Chia and Nandi[7] in relation to the electric dipole moment of the neutron.

Their two loop result for the neutron is

$$d_{neutron} \sim em_q \frac{\alpha_s g^4}{16\pi^5} \tilde{f}_{kM} (m^2_{\tilde{t}} - m^2_{\tilde{c}}) \frac{(\Delta\tilde{m}^2)^3}{\tilde{M}^8 M_W{}^2}$$

$$\sim 10^{-28} \text{ e-cm.}$$

where $\tilde{f}_{KM}$ symbolizes the pertinent combination of mixing angles and phases, and has been taken numerically equal to the usual Kobayashi-Maskawa factors for illustration purposes; $m_{\tilde{q}}$ stands for the mass of the scalar partner of the quark q; $\Delta\tilde{m}^2$ stands for intergenerational mass splittings of scalar quarks ($m_{\tilde{c}}{}^2 - m_{\tilde{u}}{}^2$, etc.) and again $\tilde{M}$ ~

1 TeV. We can extend this to the case of the electron[6], and obtain

$$d_{electron} \sim 10^{-32} \text{ e-cm.},$$

again taking $f_{KM} \sim 10^{-5}$ for leptons.

In addition, if a mixing between the scalar quarks and scalar leptons exists of the kind considered earlier (second possibility), the necessary helicity flip may be done in the internal scalar quark line. This enhances the effect so much that from the present experimental bounds for both the electron and neutron, very interesting (although model dependent) constraints are obtained[6] for products of mixing angles, mass differences in the scalar quark and lepton sectors and inverse powers of the SUSY breaking mass.

Of course, all of these sources of CP-violation also give contributions to ε'/ε. It appears[8] that in most models a CP-violation "à la Kobayashi-Maskawa" is necessary, at least for the hadrons: if the only source of CP-violation would be any of the first two mechanisms pointed out here, the value of ε and the actual bounds on ε'/ε and the neutron electric dipole moment would be incompatible, in the simplest cases.

We conclude that a dipole electric moment for the electron within one or two orders of magnitude of the value for the neutron is a natural possibility in many supersymmetric models, contrary to the situation in the standard $SU(2)_L \times U(1)$ models. Both could be observed near the actual experimental bounds.

A similar situation is present in the framework of any family symmetry, if this is the only source of CP-violation. However, there we can only expect a bound of

$$d_{electron} < 10^{-27} \text{ e-cm.}$$

and possibly $d_e \sim 10^{-30}$ e-cm. (The neutral sector of Higgs' models of CP-violation can induce also sizeable $d_{neutron} \sim d_{electron}$.) We thus conclude that any improvement of these bounds will impose interesting constraints on the parameters of SUSY models and might help in discriminating among different CP-violation mechanisms.

REFERENCES

1. "Specifically Supersymmetric contribution to electric dipole moments", F. del Águila, M.B. Gavela, J.A. Grifols and A. Méndez, Brandeis preprint, February 1983.
2. See, for instance; J. Ellis, L. Ibanez and G.G. Ross, Ref. TH-3382-CERN (1982).
3. W. Buchmüller and D. Wyler, Phys. Lett. 121B, 321 (1983).
4. J. Ellis, D. Nanopoulos and K. Tamvakis, Phys. Lett. 121B, 123 (1983).
5. J. Ellis, S. Ferrara and D.V. Nanopoulos, Phys. Lett. 114B, 231 (1982).

6. F. del Águila, M.B. Gavela, J.A. Grifols and A. Méndez, in preparation.
7. S.P. Chia and S. Nandi, Phys. Lett. 117B, 45 (1982).
8. J.M. Frere and M.B. Gavela, in preparation.
9. M.B. Gavela and H. Georgi, Phys. Lett. 119B, 141 (1982).
10. H.Y. Cheng, preprint PURD-TH-83-3.

RARE K and μ DECAYS: THEORETICAL AND EXPERIMENTAL STATUS AND PROSPECTS

Gordon L. Kane
Physics Department, University of Michigan, Ann Arbor, MI 48104

and

Robert E. Shrock
Institute for Theoretical Physics, State University of New York at Stony Brook, Stony Brook, N.Y., 11794

I. THEORETICAL CONSIDERATIONS

At present there is a remarkably successful standard model of electroweak and strong interactions. There are no contradictions between experiment and theory! Nevertheless, there are many questions which the standard model does not answer, questions of the "why" kind. The origin of quark and lepton masses is not understood, nor is the reason for generations , or how many of these there are. In the most fundamental sense, it is hoped that rare decays will occur that are not required by the standard model (that is, their rates will be nonzero if predicted to be zero in the standard model, or will differ significantly from a nonzero standard model prediction). If indeed this happens, one may expect that it will provide important clues to help answer the open questions.

There are two classes of decays. Some are expected to occur at some level in the standard model (henceforth denoted as SM), and provide tests of this model. Some examples of these are given in Table 1. New clues could arise from finding rates or other observables which differ from those predicted by the SM. For the second type of decays, the SM prediction is zero--they do not occur at any level. Finding them at all would demonstrate the existence of new interactions. For the various processes to be discussed below, these SM predictions will be mentioned; they are summarized in the first line of Table 1.

Some of the decays might give results different from the SM because of the existence of new interactions, or equivalently, the exchange of new particles. Others might be of interest as a way to discover neutrino masses because of modified kinematical distributions, or a decay into a new kind of particle.

In this section we shall provide a brief description of some ideas that might lead to observable new effects. While many of the ideas <u>suggest</u> that new effects should occur, they do not <u>require</u> that they occur at any specified level. Sometimes particular models may predict an effect, but at the present level of understanding, one can usually find another model which predicts a smaller or larger effect. Rather than trying to provide complete and detailed references in the proceedings of a workshop such as this, we shall only mention a few reviews from which the literature can be traced.[1] Partly this is because of the perspective that the ideas discussed do provide good motivation that some effects should occur somewhere, but no particular idea should be interpreted to imply that a specific prediction (beyond the SM) is more than suggestive.

TABLE I

Decay or Process / Motivation & Physics Issue & Process	$K^+ \to \pi^+ \mu^+ e^+$ $K_L^o \to \mu^\pm e^\mp$ (II)	$K_L^o \to \begin{cases} \mu^+\mu^- \\ e^+e^- \end{cases}$ (II)	$K^+ \to \begin{cases} \pi^+ e^+ e^- \\ \pi^+ \mu^+ \mu^- \end{cases}$ (II)	$K^+ \to \pi^+$ + Missing Neutral(s) (III)	$K^+ \to \begin{cases} \mu^+ \nu_i \\ e^+ \nu_i \end{cases}$ $\pi^+ \to \begin{cases} \mu^+ \nu_i \\ e^+ \nu_i \end{cases}$ ν_i Heavy, Subdominantly coupled
1) Predicted by Standard Model	N	Y	Y	$\nu\bar{\nu}$: Y other: N	N
2) ν Masses, Heavy Leptons	negligible unless very heavy	N	N	Y, P	Y
3) Lepton family number violation	Y	N	N	N	N
Total Lepton number violation	N	N	N	N	N
4) e−μ Universality violation	N	Y	Y	N	Y
5) Horizontal Interactions	Y	Y	Y	Y, P	N
6) Leptoquark Interactions	Y	Y	Y	Y	N
7) Flavor-changing Higgs	Y	Y, P	Y,P	Y,P	Y
8) Technicolor	Y	Y			
9) Supersymmetry	Y	Y, P	Y,P	Y,P	N
10) Axions,	N	N	N	Y	N
11) Composite Models process	Y	N	N	N	N

Y(N): Physics of type (1)-(11) causes or influences (does not cause or significantly influence) the given process

P : While caused or influenced by the physics of type (1)-(11), process constitutes only a poor probe for it

TABLE I (continued)

	$\pi^+ \to \mu^+ \nu_2$ $K_L^0 \to \pi^\pm \mu^\mp \overset{(-)}{\nu}_2$ $K^+ \to \pi^0 \mu^+ \nu_2$ (V)	$K_L^0 \to K^\pm e^\mp \overset{(-)}{\nu}_\ell$ (VII)	$K^+ \to \ell^+ \nu_\ell \gamma$ $\pi^+ \to \ell^+ \nu_\ell \gamma$ $\ell + e, \mu$ (VII)	$K^+ \to \pi^+ \gamma\gamma$ $K_L^0 \to \gamma\gamma$ (VII)	a $\pi^+ \to \pi^0 e^+ \nu_e$ b $\pi^0 \to e^+ e^-$ c $\pi^0 \to \gamma\gamma\gamma$ d $\pi^0 \to$ Missing Neutrals (VII)	Regular decay ρ, η, ξ, δ (VI), (VIII)	$\mu^+ \to e^+ \gamma$ $\mu^+ \to e^+ e^+ e^-$ $\mu^+ \to e^+ \gamma\gamma$ $\mu^- N \to e^\pm N'$ (VIII)
1)	Y, for $m(\nu_2)=0$	Y	Y	Y	a,b :Y; c,d: N	Y	N
2)	Y	N	N	N	a,b,c: N; d: Y	Y	Y if very heavy
3)	N	N	N	N	N	N	Y
4)	Y	N	N	N	N	Y	Y
5)	N	N	N	N	a,c: N, b,d: Y	Y	Y
6)	N	N	N	N	N	N	N
7)	Y, P	N	N	N	a,Y,P; b,c,d:N	Y	Y
8)	N	N	N	N	N	N	Y
9)	N	N	N	N	a,b,c:N:d:Y,P	Y	Y
10)	N	N	N	N	N	N	N
11)	N	N	N	N	N	N	Y

At the most fundamental level, one must recognize that there is no understanding of the fermion mass spectrum. Such an understanding would, for example, entail an ab initio calculation of the mass ratio m_e/m_μ, ideally to an accuracy comparable with the accuracy of 3 parts per million with which it is known or, at least, to a few per cent. Instead, all one has is results such as the relation[2] $m_e/m_\mu = m_d/m_s$, which is not based on any calculation, but rather on an ad hoc choice of Higgs representations within a specific GUT group, SU(5), and in any case is wrong by an order of magnitude! In the absence of such an understanding, it is not valid to infer that neutrinos have zero masses, just because their upper mass limits are small compared to the masses of their corresponding charged leptons. Indeed, for at least one "known" neutrino, ν_τ, the present upper limit on its mass, viz., 250 MeV, is not even small on a particle physics scale. There are two points of view that would actually lead one to suspect that neutrino masses may be nonzero. First, in the context of gauge theories, it is expected that a symmetry is needed to have a zero-mass particle. Such a symmetry could be a gauge or chiral symmetry, or a spontaneously broken (continuous) global symmetry. None of these necessarily applies to neutrinos. Secondly, sometimes models suggest that neutrino masses should be nonzero-- one gets such effects in some grand unified theories, some horizontal symmetry theories, and in various other models. Finally, of course, one never measures any mass to be exactly zero, and, as noted above, the upper limit on at least one neutrino mass is not very small.

Proceeding to another area of theoretical model building, we note that there is no understanding at all of the meaning of flavor. If the lesson of the past decade that one should try to interpret particle physics in terms of gauge theories is relevant, then a local (or conceivably, a global) horizontal symmetry may be applicable. Then, horizontal gauge bosons or Higgs bosons, or possibly Goldstone bosons, may cause flavor transitions such as $s \leftrightarrow d$, $c \leftrightarrow u$, $b \leftrightarrow s$, $t \leftrightarrow c$, $\mu \leftrightarrow e$, $\tau \leftrightarrow e$, $\tau \leftrightarrow \mu$, etc. These could induce decays that are zero in the SM, or modify the rates for other decays.

In the SM, the SU(2)×U(1) symmetry is broken, and masses are introduced, by coupling new scalar bosons (the Higgs bosons) to gauge bosons and fermions. The ad hoc nature of this procedure, and the difficulty of dealing with fundamental scalar bosons, whose masses are normally sensitive to the highest mass scale present in the theory, has led to several new approaches, particularly to technicolor and supersymmetry models. (Supersymmetry is a possibility independent of this problem and, in its local realization as a gauge symmetry, may help one to understand quantum gravity, but some of the present interest in it, at least among particle physicists, derives from the above-mentioned connection.) Technicolor basically assumes that rather than having fundamental scalar bosons, one should introduce new fundamental fermions and a new gauge interaction (viz., techicolor) analogous to color QCD. Then the bound state "pions" of the new force are the "Higgs bosons" of the SM. The mechanism by which gauge bosons get mass is different from that for fermions in this approach, the latter requiring yet another interaction with massive bosons, namely the so-called "extended technicolor" or ETC interaction.

These new bosons have masses which are constrained to give the known fermion masses. At the same time, the new bosons will couple different flavors together, unless symmetry constraints are imposed, thus potentially yielding flavor-changing decays. Since masses of the new bosons are not free parameters, one can estimate the expected rates for decays such as $K_L^o \to \mu^\pm e^\mp$, and typically, one gets rates of the order of (or, indeed, larger than) present limits. Technicolor theories include as bound states both neutral pseudoscalar bosons and leptoquark bosons that can also mediate neutral flavor-changing transitions. The rates expected from both of these sources are again at about the level of the experimental limits. While it could happen that symmetries reduce the expected rates considerably, the general interpretation is that substantial motivation for anticipating rare decays is provided by technicolor ideas, particularly since the mass scales are not very flexible in such theories.

A different approach to dealing with scalar (Higgs) bosons is that of supersymmetry, where the scalars are just as fundamental as the fermions. One should recognize that, despite some attempts to hybridize supersymmetry and technicolor (e.g., supercolor), the basic philosophies underlying the two theories are fundamentally antithetical. The uneasy coexistence of both approaches in the present mélange of theoretical speculations affords some evidence of just to what extent these speculations are simply groping in the dark. From the point of view of rare decays, there are two kinds of possible implications. In supersymmetric theories, every known particle has a partner differing in spin by ½ unit. Needless to say, none of these additional particles has been observed in any existing data. Some of the particles might be light (e.g., the photino, $\tilde{\gamma}$, the partner of the photon) and occur as final state particles in various decays such as $K^+ \to \pi^+\tilde{\gamma}\tilde{\gamma}$. Alternatively, supersymmetric partners might occur internally in Feynman diagrams and thereby generate additions to several of the rare decays. As always, this may be avoided or highly suppressed by imposing certain relations among masses and/or couplings, but given the mass scales expected in supersymmetric models, it would not be surprising if rare decays were induced.

Supersymmetric and technicolor approaches are, as emphasized above, quite different ones since the former treats scalars as fundamental while the latter treats them as composite. If rare decays are found, it may be possible to distinguish between these approaches (and others to be mentioned) by studying the pattern of induced decays, the form of the induced decay currents, and other observables.

If axions or Goldstone bosons are produced in any model, they can occur as external particles X in the decay $K^+ \to \pi^+X$. This decay would give a two-body peak in what would otherwise be a three-body decay distribution (aside from background from $K^+ \to \pi^+\pi^0$). The normal axion, originally introduced as part of an effort to eliminate strong CP violation, would already have been observed in this decay if it had the standard Peccei-Quinn-Weinberg-Wilczek couplings and mass. Its coupling can be reduced in various models, such as the "invisible" axion models. Speculative possibilities such as these, or Goldstone bosons that couple one flavor to another (see the talk by Wilczek) could occur.

The proliferation of supposedly fundamental quarks and leptons, and the inability of current theory to provide any explanation or understanding of the masses of these particles, has led many physicists to question whether indeed these fermions are fundamental, and to investigate the implications of their possible composite structure. This line of reasoning is very natural in view of the historical development of physics, in which, as one probed to higher and higher energies and shorter and shorter distances, level after level of purportedly fundamental matter was found to be composite. The false conclusions of generations long dead are even enshrined in common terminology -- classic examples are Democritus' "atom" from the Greek "ατομος", meaning indivisible, and "proton" from "πρωτος" meaning first or most fundamental. It is a staggering hubris indeed (although presently consistent with all available data) to assume that somehow we are more fortunate than all our ancestors, that we have reached the final, most basic, interactions and constituents of matter, beyond which there are no others. This attitude is exemplified by the famous desert hypothesis, according to which there is no new physics between $\sim 10^2$ GeV and the grand unification scale of $\sim 10^{14}$ GeV. There are several mechanisms in composite models which can induce rare decays. In such models one can rearrange constituents and get bosons with leptoquark quantum numbers.[3] These can include vector bosons whose couplings are not suppressed by factors of the form $(m_{fermion}/m_{vector\ boson})$ and whose exchange could induce rare decays at interesting levels, unless certain symmetries are imposed. Excited q^* or ℓ^* states can occur in loop diagrams with non-diagonal couplings and hence produce rare decays. Constituent rearrangement can occur and yield effective four-fermion neutral, flavor-changing interactions.

In the above remarks we have provided a short introduction to some of the motivations for searching for rare decays. It should be remembered that the basic motivation is simply to find new interactions in nature, to help integrate masses and flavors into the theory, and to understand why the standard model takes the form that it does. We next proceed to give more detailed discussions of specific decays.

II. THE DECAYS $K^+ \to \pi^+\mu^+e^-$ and $K_L^0 \to \mu^\pm e^\mp$

These decays are of interest because they violate lepton family number and can probe possible new physics in the multi-TeV mass range. The present upper bounds on their branching ratios are[4]

$$B(K^+ \to \pi^+\mu^+e^-) < 5 \times 10^{-9} \quad (90\%\ CL) \tag{2.1}$$

and

$$B(K_L^0 \to \mu^\pm e^\mp) < 6 \times 10^{-6} \quad (90\%\ CL) \tag{2.2}$$

(In the latter case a more stringent bound $B(K_L^o \rightarrow \mu^{\pm}e^{\mp}) < 2 \times 10^{-9}$ was reported by an old LBL experiment[4] but is not included in the Particle Data Tables because of uncertain systematic errors in the experiment.) The limit

$$B(K^+ \rightarrow \pi^+\mu^-e^+) < 7 \times 10^{-9} \text{ (90\% CL)} \tag{2.3}$$

has also been achieved.

There are now two experiments approved to search for the decays (2.1) and (2.2) at the Brookhaven AGS. A Yale-BNL-Seattle collaboration (E777, M. Zeller, spokesman) will search for the decay $K^+ \rightarrow \pi^+\mu^-e^+$ with an estimated sensitivity of $\sim 10^{-11}$ in branching ratio.[5] It will use a 6 GeV/c beam of K^+'s and a detector with a gas Čerenkov counter, e^+ calorimeter, μ^+ range counter, and very good particle identification. The main background is from $K^+ \rightarrow \pi^+\pi^+\pi^-$ where one of the π^+'s decays to $\mu^+\nu_\mu$ and the π^- is misidentified as an e^-. This background is expected to produce spurious events at the level of a few $\sim 10^{-12}$ in branching ratio. Secondly, a Yale-BNL collaboration (E780, M. Schmidt and W. Morse, cospokesmen) will search for the decay $K_L^o \rightarrow \mu^+ e^-$ down to an anticipated level of 10^{-10} in branching ratio.[6,7] This group utilizes minidrift chambers with $\sim 200\,\mu$ resolution. The main background is from $K_L^o \rightarrow \pi^+e^-\bar{\nu}_e$ where the π^+ decays to $\mu^+\nu_\mu$ in the spectrometer magnet and comes in at a level $\sim 2 \times 10^{-11}$ in branching ratio. Both of these experiments will be on line in early 1985, or possibly earlier, and should have first results by early 1986. They would both benefit from the cleaner environments which would be possible with more intense beams, and it has been roughly estimated[7] that with requisite improvements in detectors, one might be able to achieve sensitivities of 10^{-12} for $B(K^+ \rightarrow \pi^+\mu^+e^-)$ and 10^{-11} for $B(K_L^o \rightarrow \mu^+e^-)$ if such experiments were performed with better beams.

There are a number of models that could give rise to effective leptonic flavor-changing neutral current (LFCNC) decays such as these. These include:

(1) an extended electroweak theory with flavor-changing Higgs bosons; (2) (extended) technicolor; (3) horizontal generation-changing interactions; (4) baryon - and lepton - violating low-mass Higgs in grand unified theories (GUT's); (5) some supersymmetric GUT's; (6) preon models; and (7) theories with very heavy neutral leptons. The theoretical motivations for these various models have been briefly described above. The contributions of the decay mechanisms are, of course, constrained by the requirement that they do not produce unacceptably large $K^o - \bar{K}^o$ mixing, $K_L^o \rightarrow \mu^+\mu^-$ decay rate, and so forth for known processes. As a rough indication of the type of limit that one might set, consider a theory with horizontal gauge interactions, involving V or V-A couplings to fermions. Let $m(V_h)$ and g_{Vh} denote the mass and gauge coupling constant in such a theory. Further, denote the gauge coupling of the SU(2) factor in the standard model as g. Then in the absence of GIM-type cancellations, one obtains the lower limits[8]

$$m(V_h) \gtrsim 20 \text{ TeV} \left(\frac{5 \times 10^{-9}}{B(K^+ \to \pi^+ \mu^+ e^-)} \right)^{1/4} \left| \frac{g_{V_h}}{g} \right| \quad (2.4)$$

and

$$m(V_h) \gtrsim 38 \text{ TeV} \left(\frac{2 \times 10^{-9}}{B(K_L^o \to \mu^+ e^-)} \right)^{1/4} \left| \frac{g_{V_h}}{g} \right| \quad (2.5)$$

Thus, if the two current experiments at BNL do not detect any signals down to their anticipated levels of sensitivity, $B(K^+ \to \pi^+\mu^+ e^-) \sim 10^{-11}$ and $B(K_L^o \to \mu^+e^-) \sim 10^{-10}$, one may infer the bounds $m(V_h) \gtrsim 90$ TeV and $\gtrsim 80$ TeV, respectively. It should be stressed that these limits are highly model-dependent and are given only as a rough indication of the scale of vector boson masses that may be probed within the next few years. For example, if the horizontal gauge bosons coupled in a purely vectorial manner to quarks, then they would not contribute to the decay $K_L^o \to \mu^+e^-$, which involves only an axial-vector hadronic current. On the other hand, if these horizontal gauge bosons coupled in a purely axial-vector manner to the quarks, then they would not contribute to the decay $K^+ \to \pi^+\mu^+e^-$, which proceeds via the vector hadronic current. In passing, we note that in this class of theories, the $|\Delta G| = 2$ decay $K^+ \to \pi^+e^+\mu^-$ would be suppressed relative to the $\Delta G = 0$ decay $K^+ \to \pi^+\mu^+e^-$, (Here, G denotes generation, with $G = 1$ for $\{u,d,\nu_e, e\}$, $G = 2$ for $\{c,s,\nu_\mu,\mu\}$, etc.) Both of the decays $K_L^o \to \mu^+e^-$ and $K_L^o \to \mu^-e^+$ contain $\Delta G = 0$ parts as well as $|\Delta G| = 2$ parts. The possibility that there may be such horizontal gauge interactions mediated by vector bosons with masses in the range accessible to the current BNL experiments, viz., ~ 100 TeV (for typical theories with $g \sim g_{Vh}$), is an intriguing one. However, there is no particular reason to expect that this situation actually obtains in nature. Similarly, there is no particular reason why effective LFCNC interactions arising in SSGUT's or preon models should be characterized by a mass scale of order 1-100 TeV. Very heavy neutral leptons with masses greater than $\sim$ 1GeV, which might couple virtually in loop diagrams contributing to these rare K decays are certainly possible, but again, there is no strong motivation for supposing that they actually exist. (Of course, before the discovery of the muon there would not have been any theoretical motivation for supposing that it existed either, and there is still no theoretical understanding of why it and the other fermions of the second and third generations exist.) Moreover, heavy neutrinos with masses $\lesssim$ 350 MeV are constrained to have very small couplings to e and μ.[9,10] There are other indirect constraints on the couplings of heavy neutrinos with masses above m_K which again severely limit their contribution to LFC K decays.[11,12] Thus, it is unlikely (although possible) that category (7) would yield leptonic flavor-changing K decays at levels which are experimentally accessible at a foreseeable kaon factory.

In contrast, the TeV mass scale is probably a natural one for Higgs bosons in electroweak theories, and, in particular, flavor-changing Higgs. Similarly, the effective FCNC interactions resulting from technicolor models are typically characterized by mass scales of order 10-100 TeV.[13-15] We suspect however, that specific claims such as the claim[15] that "If at the quark vertex the ETC interactions are not purely vectorial, as is expected because of the u-d mass splittings, then the decay $K_L^0 \to \mu^+ e^-$ should probably show up within an order of magnitude below the present experimental limit. If the ETC interactions are not purely axial at the quark vertex, then the decay $K^+ \to \pi^+\mu^+e^-$ should probably show up within two orders of magnitude of the present upper limit," are somewhat too strong because of the severe model dependence involved, and the very speculative nature of the models themselves. Unfortunately, the models being discussed here are considerably less well-formed, definite, and predictive than the standard electroweak theory. Thus, there are not yet firm predictions that such-and-such a decay will occur at a given level in branching ratio. The situation is, rather, that there are theoretical motivations for considering a number of various models which go beyond the standard "low" -energy SU(3)×SU(2)×U(1) theory; and within certain of these models, FCNC decays such as the two possible K decays under discussion here may occur at rates which are within range of planned or future experiments.

An experiment that searches for these rare decays can also measure and/or search for other related conventional decays. Specifically, an experiment looking for $K^+ \to \pi^+\mu^+e^-$ should obtain a good sample of $K^+ \to \pi^+e^+e^-$ events, since $B(K^+ \to \pi^+e^+e^-) = (2.7 \pm 0.5) \times 10^{-7}$, four orders of magnitude greater than the expected sensitivity of the experiment. This large sample would enable one to study the e^+e^- invariant mass spectrum and search for peaks such as would be due to $K^+ \to \pi^+ H$; $H \to e^+e^-$, where H denotes a generic Higgs boson. It is true that the "natural" value for the Higgs mass is one comparable to its vacuum expectation value, ∿ 250 GeV, but light Higgs are still possible, at a phenomenological level. This study of the e^+e^- spectrum could also yield some further information concerning the decay mechanism.[16] In addition, one could search for the decay mode $K^+ \to \pi^+\mu^+\mu^-$, for which there is only the upper limit $B(K^+ \to \pi^+\mu^+\mu^-) < 2.4 \times 10^{-6}$. This decay will certainly occur via one-loop electroweak diagrams involving emission of a virtual photon which creates the $\mu^+\mu^-$ pair.[16] The branching ratio is suppressed significantly relative to that for the $K^+ \to \pi^+e^+e^-$ mode because of the reduced phase space available. (However, there is no photon propagator suppression, because the $1/q^2$ from this propagator is cancelled by the q^2 from the quark part of the diagram, which consists dominantly of a nondiagonal charge-radius term. A priori, one would also be interested in examining the $\mu^+\mu^-$ invariant mass distribution to search for a Higgs boson which was massive enough to decay to a $\mu^+\mu^-$ pair. However, the existence of a massive Higgs boson with standard couplings to fermions has been ruled out for $m_H < 409$ MeV by a recent experiment on the decay $\eta' \to \eta\mu^+\mu^-$ which made the analogous search for a peak in the $\mu^+\mu^-$

invariant mass distribution.[18] This result forbids the decay $K^+ \to \pi^+ H$; $H \to \mu^+\mu^-$. Concerning present experiments, the Yale-BNL group does plan to study the $K^+ \to \pi^+ e^+ e^-$ decay mode and measure the e^+e^- invariant mass spectrum from 140 MeV to 350 MeV. (It does not plan to search for the decay $K^+ \to \pi^+\mu^+\mu^-$ because of the very severe background from the decay $K^+ \to \pi^+\pi^+\pi^-$, where a $\pi^+\pi^-$ pair is misidentified as $\mu^+\mu^-$.)[5]

Similarly, an experiment which searches for the decay $K_L^0 \to \mu^\pm e^\mp$ can amass a sizeable sample of $K_L^0 \to \mu^+\mu^-$ events. This decay has long been of interest as a higher-order induced neutral $|\Delta S| \neq 0$ weak process.[19] Like K^0-$\bar{K}^0$ mixing and other one-loop induced rare K decays, it is sensitive to the couplings of heavy quarks. In particular it provides constraints on the quark mixing matrix coefficients that determine the strength of the charged current couplings of these heavy quarks to d and s.[20] The present data on this decay comes from the three experiments by Columbia-BNL[21] , Princeton-BNL[22], and Chicago-FNAL[23] groups and consists of 27 events. A sample of $\sim 10^3$ events ought to be obtainable in a current experiment. This would obviously yield a commensurately more accurate measurement of the branching ratio and, in addition, could render feasible a measurement of the muon polarization. The latter might be of some interest as a probe of the decay mechanism: if the conventional two-photon intermediate state and similar long-distance contributions really dominate the amplitude,[24] then this polarization, $P_\mu = 0$. This is also true for the part of the amplitude arising from the short-distance, quasi-free quark diagrams.[25] However, it is conceivable that FCNC interactions with effective S,P Lorentz structure could contribute significantly to $K_L^0 \to \mu^+\mu^-$ while not giving too large a rate for $K^0 - K^0$ mixing. This would require rather artificial fine tuning of certain couplings (the coupling to eμ would have to be much larger than the coupling to sd), but is nonetheless possible. Through interference effects, one can then obtain a nonzero muon polarization.[26] Although unlikely, it is possible that this might be large enough to measure in a dedicated experiment.

An experiment on $K_L^0 \to \mu^+ e^-$ could also search for the decay $K_L^0 \to e^+e^-$, for which the present limit upper limit is $B(K_L^0 \to e^+e^-) < 2.0 \times 10^{-7}$ (90% CL). To estimate a lower limit on the branching ratio for this mode, one may simply take the contribution of the two-photon intermediate state to the absorptive part of the amplitude, which gives[24]

$$\frac{B(K_L^0 \to e^+e^-)_{2\gamma\ \mathrm{abs}}}{B(K_L^0 \to \mu^+\mu^-)_{2\gamma\ \mathrm{abs}}} \simeq \frac{3 \times 10^{-12}}{6 \times 10^{-9}} \qquad (2.6)$$

The real part of the amplitude is harder to estimate. The short distance quasi-free quark contributions to both the real and imaginary parts of the amplitude, in the standard model, give rates which scale like

$$\frac{B(K_L^o \to e^+e^-)_{f.q.}}{B(K_L^o \to \mu^+\mu^-)_{f.q.}} = \frac{m_e^2}{m_\mu^2}\left(\frac{1-\frac{4m_e^2}{m_\pi^2}}{1-\frac{4m_\mu^2}{m_\pi^2}}\right)^2 = 3.5\times 10^{-5} \tag{2.7}$$

Long-distance contributions to the real part are also present at some level, although arguments have been given that they are small.[25] The suppression factor (2.6) is due to the vector nature of the electromagnetic couplings in the $2\gamma_{virtual}$ diagram, while that in (2.7) is due to the fact that the quasi-free quark amplitude takes the form

$$\{<0|\bar{s}_L\gamma_\lambda d_L|K^o> + <0|\bar{d}_L\gamma_\lambda s_L|\bar{K}^o>\}\ [\bar{u}_\mu\gamma_\lambda\ (a + b\gamma_5)v_\mu]$$

$$\sim \text{const.}\ \ 2m_\mu\ b\ \ [\bar{u}_\mu\gamma_5 v_\mu]\ . \tag{2.8}$$

Here the s-channel Z-exchange and WW box graphs contribute to both a and b in the quark amplitude, but the s-channel γ-exchange contributes only to a. As the a term does not enter in the physical amplitude, the effective lepton current is purely axial-vector. If one assumes that this effective (V,A) leptonic current also applies to the real part, then

$$B(K_L^o \to e^+e^-) = \frac{B(K_L^o \to \mu^+\mu^-)}{B(K_L^o \to \mu^+\mu^-)_{2\gamma\ abs}} \times B(K_L^o \to e^+e^-)_{2\gamma\ abs}$$

$$\simeq 5\times 10^{-12} \tag{2.9}$$

This is below the level of 10^{-10} in branching ratio to which the present Yale-BNL experiment is statistically sensitive. However, precisely because of this standard-model suppression of the ee mode, it serves as a good probe for nonstandard contributions. For example, if there is an effective FCNC interaction with (S,P) Lorentz structure, then the size of the $e\bar{e}$ mode might be increased, relative to that of the $\mu\bar{\mu}$ mode. It should be noted, however, that if this interaction arises directly from Higgs exchange, then the $(m_e/m_\mu)^2$ suppression factor might very well be reinstated, since Higgs bosons typically (although not necessarily) couple to fermions with strengths $\sim (m_f/m_V)$, where m_V denotes a generic vector boson mass in the theory. Thus, even if such additional interactions were present, they might well be suppressed for the $e\bar{e}$ mode in a manner much like that in the standard model case. Nevertheless, it is clearly possible and very worthwhile to search for this decay; a positive signal at a level significantly higher than that expected in the standard model would indicate new physics, while a null result down to $B(K_L^0 \to e^+e^-) \sim 10^{-10}$ would still constitute a great improvement over the existing upper limit.

III. THE DECAY $K^+ \to \pi^+$ + MISSING NEUTRAL(S)

(A) The Case of One Missing Neutral Particle: $K^+ \to \pi^+ X^0$

This subsumes the actual decays $K^+ \to \pi^+ H$, $\pi^+ a$, or $\pi^+ f$, where H, a, and f denote a Higgs boson, axion,[27] and familon,[28] respectively, as well as any other generic neutral spin-0 boson. It is necessary that the X^0 not decay in a visible manner in order for the observed signal to be of this type. A recent experiment at KEK has set the very stringent upper limit[29]

$$B(K^+ \to \pi^+ + \text{axion}) < 3.8 \times 10^{-8} \quad (90\%\ \text{CL}) \tag{3.1}$$

for $m_a << m_\pi$. This upper limit is substantially smaller than most calculations of the branching ratio due to the standard (light) axion and almost certainly rules it out.[30] The mass of the standard axion is[27]

$$m_a \sim 150\ \text{KeV} \left(\frac{n}{3}\right) \left(\frac{x + x^{-1}}{2}\right) \tag{3.2}$$

where n denotes the number of quark generations and x denotes the ratio of vacuum expectation values of the two Higgs fields in the Peccei-Quinn model. Thus, if, as might be considered natural, x is not too far from unity, then $m_a \sim 150$ KeV. However, if $x << 1$ or $x >> 1$, then m_a might be large enough not to have produced a peak in the π^+ momentum spectrum at the position $(m_K/2)(1 - m_\pi^2/m_K^2) \simeq$ 229 MeV looked for by the experiment of Ref. 29. It might still have been detected if it decayed sufficiently rapidly to $\gamma\gamma$, since this experiment also compiled data on the decay mode $K^+ \to \pi^+\gamma\gamma$.[31] Obviously, if $m_a \simeq m_{\pi^0}$ then the $\gamma\gamma$ decay signal would be swamped by far more numerous events from $K^+ \to \pi^+\pi^0$; $\pi^0 \to \gamma\gamma$. In any case, theorists have long since invented the "invisible" axion whose coupling to matter is $\sim g(m_f/M)$ where $M \sim 10^{12}$ GeV, rather than $g(m_f/m_W)$ as for the original axion, and hence is strongly suppressed.[32]

In addition to the possibility of heavy axions, there is also the possibility of a Higgs boson in the mass range where it could be emitted in a decay of the form $K^+ \to \pi^+ H$. A more recent KEK experiment[33] has searched over a wide mass range for such a light Higgs, or more generally X^0, with no visible decay, by means of looking for the peak which it would cause in the π^+ momentum spectrum. This experiment has established a correlated upper limit on $B(K^+ \to \pi^+ X^0)$, as a function of m_{X^0}, which varies from $\sim 10^{-6}$ for $m_{X^0} \lesssim 75$ MeV to

$\sim 10^{-5}$ for 170 MeV < m_{X^o} < 260 MeV. For $m_{X^o} \sim m_{\pi^o}$ the limit obtained is less stringent because of background from the decay $K^+ \to \pi^+\pi^o$, where the two photons from the π^o decay are not detected. A Higgs boson with a mass in the range < $2m_\mu$ would normally decay dominantly to e^+e^- with a lifetime short enough to be detectable. However, one should keep in mind the possibility that for an interesting range of masses, the Higgs boson could decay primarily in an invisible way, viz., $H \to aa$, hh, MM, or $\nu\bar{\nu}$, where h and M denote a lighter Higgs and a Majoron, respectively, and ν denotes a massive neutrino.[34] Similar comments apply to a heavy axion or familon.

As with the other decays surveyed herein, the limits achieved by these KEK experiments on the decay $K^+ \to \pi^+ X^o$ involving a light or massive X^o could be improved by further detector developments and greater statistics.

(B) The Case of Several Missing Neutral Particles

The decay $K^+ \to \pi^+$ + missing neutrals would be the observed signal resulting from one definite standard-model source,[16] $\sum_{i=1}^{3} K^+ \to \pi^+ \nu_i\bar{\nu}_i$, and from a number of possible sources involving new physics. Like some other rare K decays such as $K_L^o \to K^\pm e^\mp (\bar{\nu})_e$, this mode has the advantage that since it occurs in the standard model, if one can perform an experiment with the required sensitivity, one can expect to see a signal rather than just obtaining a null result, as might happen for the lepton flavor-violating decay modes. However, it has two important disadvantages. First, given the uncertainties in the standard-model prediction, even if a signal is seen, it is not clear at what level one will be able to claim that one has seen new physics. This is in sharp contrast to the situation for the decays $K^+ \to \pi^+\mu^+e^-$ and $K_L^o \to \mu^+e^-$, for which the observation of a nonzero signal by itself constitutes new physics beyond the standard model. Secondly, even if the observed rate differs enough from the standard model so that one can convincingly make a case for new physics, it will not in general be possible to determine what this new physics is! These are significant disadvantages to weigh against the possibility of not seeing any signal in the lepton flavor-violating decay modes.

The conventional source $\sum_{i=1}^{3} K^+ \to \pi^+ \nu_i\bar{\nu}_i$, gives[16,35-38] $B(K^+ \to \pi^+ + \text{missing neutrals}) \sim$ few $\times 10^{-9}$ to 10^{-10}. There are uncertainties in the calculation having to do with the mass of the t-quark, which is currently unknown, and the associated weak couplings V_{td} and V_{ts}, for which approximate bounds have been derived. (Of course, the existence and mass of the t-quark may be established directly at PETRA or inferred indirectly from W and/or Z decay data from the CERN $\bar{p}p$ collider in the near future.) Equally important, there is an uncertainty arising from the parametrization of the q^2-dependence in the form factors for the hadronic matrix element.

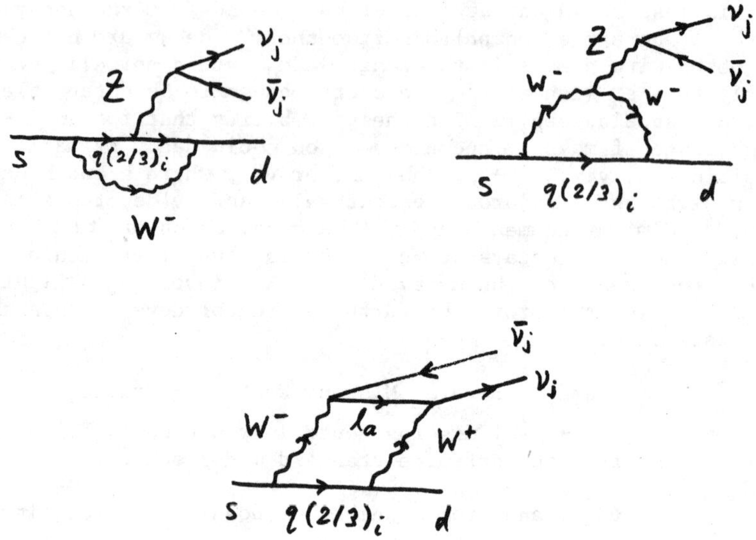

(a) Graphs for the decay $K^+ \to \pi^+ \nu_j \bar{\nu}_j$

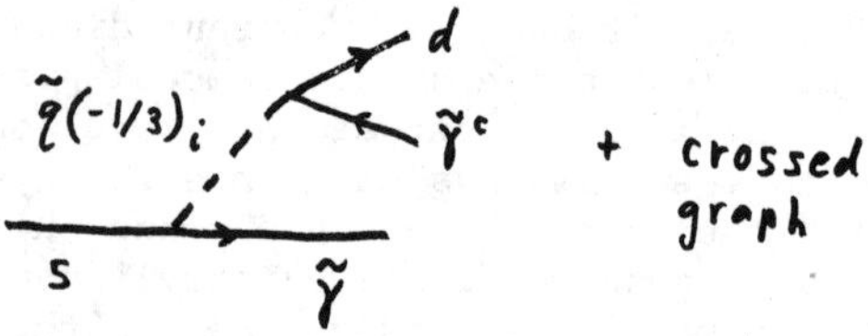

(b) Tree-level graph for $K^+ \to \pi^+ \tilde{\gamma}\tilde{\gamma}$

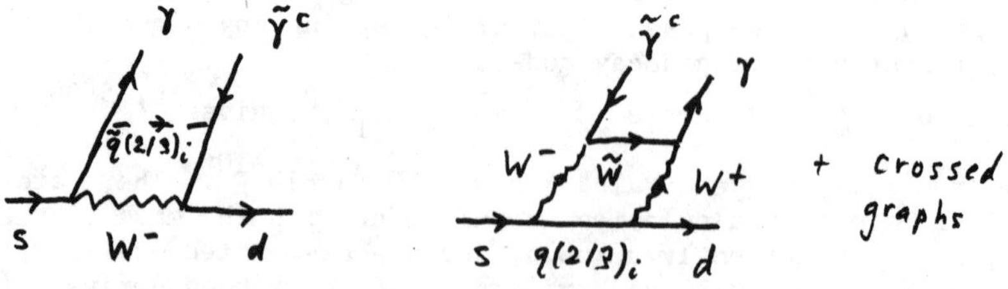

(c) Representative one-loop graphs for $K^+ \to \pi^+ \tilde{\gamma}\tilde{\gamma}$

Figure 1

$$\langle\pi^+(p_\pi)|\bar{s}_L\gamma_\lambda u_L|K^+(p_K)\rangle \sim f_+(q^2)(p_K+p_\pi)_\lambda + f_-(q^2)q_\lambda \tag{3.3}$$

where $q = p_K - p_\pi$. The decay depends, like the decay $Z \to \Sigma\, \nu_i\bar{\nu}_i$ and the reaction $e^+e^- \to \nu\bar{\nu}\gamma$, on the number of neutrino types. However, since m_t is not known at present, one does not, strictly speaking, have any prediction for the branching ratio, even for the case of $n = 3$ generations. This is significant for groups that are considering doing this experiment in the near future. Assuming that the t-quark is observed and its mass determined before a possible experiment is run, one would at least have a standard-model calculation for $\sum_{i=1}^{3} B(K^+ \to \pi^+ \nu_i\bar{\nu}_i)$, which could be expected to be accurate to within perhaps a factor of 3 either way. The problem is that, given this uncertainty, one could not perform any convincing test of the standard model prediction or test for $n > 3$ generations. First, as noted at the beginning, since one does not observe what the missing neutrals are, one obviously cannot claim that they are additional neutrinos of higher generations, nor can one claim that they are photinos, higgsinos, etc. Second, unfortunately, one cannot even test a more limited hypothesis such as the hypothesis that there are $n > 3$ generations in the standard model. The reason is that the test is circular: if n is assumed to be greater than 3, then the amplitude depends on the unknown masses and couplings of the additional higher generation quarks which enter in all the Feynman diagrams of Fig. 1 and in addition, on the masses of the corresponding charged leptons which enter in the WW box diagram. It is true that there are approximate constraints on the mixings of heavy quarks with light ones arising from the requirement that they do not produce an unacceptably large effect in other one-loop induced kaon processes such as $K^0 \leftrightarrow \bar{K}^0$, $K_L^0 \to \mu\bar{\mu}$, etc. However, since the functional forms of the amplitudes are not the same, these constraints cannot be taken over directly and applied to $K^+ \to \pi^+ \nu\bar{\nu}$. Even worse, the present process (with $n > 3$) also depends on the additional charged heavy leptons, the masses of which are unknown, except that they must be greater than ~ 18 GeV from PETRA data. Compounding the problems, there can be destructive interference, so that the rate for a decay $K^+ \to \pi^+ \nu_i\bar{\nu}_i$ can actually vanish at the one-loop level! Other things being equal, the values of the lepton masses at which such a vanishing would occur are in the range $\sim 10^2$ GeV, quite possible for higher generation fermions being considered.

In summary, the decay $K^+ \to \pi^+$ + missing neutrals can, logically, never used to make any specific claim about a particular channel, $\sum_i \nu_i\bar{\nu}_i$, since by assumption, one does not actually <u>observe</u> the $\nu_i\bar{\nu}_i$, and there could well be other decays which yield exactly the same experimental signature.[38] However, if and when the t-quark is "observed," and an effective value of m_t determined, and if the relevant quark mixing angles can be measured to sufficient accuracy, then a SM prediction will be possible for $n = 3$ generations, albeit with some intrinsic theoretical uncertainties, such as those connected with unknown q^2 - dependence of the hadronic form factors. A strong deviation from this prediction would demonstrate the existence of some new physics.

We proceed to consider specific contributions to $K^+ \to \pi^+$ + missing neutrals from sources beyond the standard model. One interesting source is $K^+ \to \pi^+\tilde{\gamma}\tilde{\gamma}$, where $\tilde{\gamma}$ denotes the supersymmetric partner to the photon, viz., the photino.[39,36,37,40] In all the known cases of states that have the same charge, spin, and color (if any), the eigenstates of broken symmetry groups are not actually mass eigenstates but rather linear combinations thereof. Examples include the Cabibbo-Kobayashi-Maskawa quark mixing, the mixing of the gauge eigenstates A_3 and B of the standard SU(2)×U(1) theory to form the mass eigenstates Z and γ, and the mixing of ω_8 and ω_1 to form ω and and ϕ in flavor SU(3) . In the case of supersymmetry, one thus is led in general to expect that there will be a similar mixing of the mass eigenstates of the scalar supersymmetric partners of the quark, viz., the quarks, to form the gauge group, and hence interaction, eigenstates. If this is the case, then there will be a tree-level decay diagram contributing to the decay $K^+ \to \pi^+\tilde{\gamma}\tilde{\gamma}$, as shown in Fig. 1 . The resulting branching ratio depends on (a) the CKM-type mixing angles of the squarks; (b) the squark masses; and (c) the degrees of suppression due to the super-GIM mechanism, which is operative. Consistently with other constraints such as $K^0 - \bar{K}^0$ mixing and $K_L^0 \to \mu\bar{\mu}$, this tree-level diagram could produce a branching ratio not far below the present experimental upper limit

$$B(\sum_i K^+ \to \pi^+ \nu_i\bar{\nu}_i) < 1.4 \times 10^{-7} \quad (90\%\ \mathrm{CL}) \qquad (3.4)$$

established by the KEK experiment. (The possibility of the such a tree-level decay was neglected in Ref. 37.) If the squark mixing is fine-tuned to be extremely small, then the tree-level decay could be suppressed so much that the main contributions would arise from various one-loop diagrams. Again, the resulting branching ratio is uncertain, since it depends on unknown masses of squarks and fermionic partners of gauge bosons. Estimates based on current ideas about where these masses might lie give, from the one-loop graphs,[37,40] $B(K^+ \to \pi^+\tilde{\gamma}\tilde{\gamma}) \sim 10^{-10}-10^{-11}$. There could also be other contributions from decays into fermionic partners of Higgs, called shiggses or Higgsinos: $K^+ \to \pi^+ + \tilde{H}\tilde{H}$.[37]

Another possibility is that the the photino is sufficiently massive that it would decay in a suitably designed detector.[36] There have been several analyses of astrophysical and cosmological bounds on photino masses, including one which yielded the constraints $m_{\tilde{\gamma}} < 30$ eV or $m_{\tilde{\gamma}} \gtrsim 0.3$ MeV [41] and more recently another yielding (with somewhat different theoretical inputs) the lower limit

$m_{\tilde{\gamma}} \gtrsim$ 2 -30 GeV, where the range depends on the values of squark masses. If one accepts the latter analysis, then if the $K^+ \to \pi^+ \tilde{\gamma}\tilde{\gamma}$ decay occurs at all, the photino must be so light that it will not decay visibly.[43]

IV SEARCHES FOR MASSIVE NEUTRINOS AND LEPTON MIXING IN $K_{\ell 2}$ AND $\pi_{\ell 2}$ DECAYS

In 1980 a new class of correlated tests for neutrino masses and mixing was pointed out.[12] The basic observation underlying this new class of tests was that, in general, a decay such as $\pi^+ \to \mu^+ \nu_\mu$ does not just yield a μ^+ and possibly massive ν_μ, as had previously been assumed in searches for a nonzero ν_μ "mass." Rather, if one entertains the possibility of massive neutrinos at all, as, of course, one must in testing for them, then it is not justified to assume that the weak eigenstates ν_e ν_μ ν_τ, etc. coincide with the mass eigenstates ν_1,ν_2, ν_3, etc; instead, the former are linear combinations of the latter as given by Eq. 1.2, of reference 12. It follows that a decay such as $\pi^+ \to \mu^+ \nu_\mu$, which had been used to set the upper bound on "$m(\nu_\mu)$", does not yield just a μ^+ and ν_μ, but rather consists of the sum of separate decays $\pi^+ \to \mu^+ \nu_i$ into all the mass eigenstates ν_i comprising the weak eigenstate ν_μ and allowed phase space to occur in the decay. Thus, the conventional test for "$m(\nu_\mu)$" $\neq 0$, viz., a shift in the peak in $dN/d|\vec{p}_\mu|$ downward slightly from the "$m(\nu_\mu)$" = 0 value, might very well have missed a positive signal; in the dominantly coupled mode, $\pi^+ \to \mu^+ \nu_2$, $m(\nu_2)$ might be sufficiently small that there would be no observable shift in the main peak, while a heavy, subdominantly coupled ν_i might be emitted, in the decay mode $\pi^+ \to \mu^+ \nu_i$ and might produce a peak substantially below the main one. Since previous experiments had regarded ν_μ as a mass eigenstate, they never searched for a multitude of peaks and, indeed, set cuts which would have excluded such peaks from their data even if the latter had been present. It was thus proposed[12] that new experiments be performed with $\pi_{\ell 2}$ and $K_{\ell 2}$ decays to search for possible additional peaks in the charged lepton momentum or energy spectra which would accompany the emission of heavy neutrino(s).

The peak search test is very sensitive because the signal, if it exists at all, is monochromatic and can be distinguished well from various backgrounds, which are continuous. If an additionalpeak is discovered, one can, of course, immediately determine $m(\nu_i)$ for the corresponding decay mode $M^+ \to \ell^+ \nu_i$, where $M = \pi$ or K, and, using the assumption of standard V-A couplings, one can determine $|U_{ai}|^2$, where $a = 1$ or 2 for $\ell = e$ or μ, respectively. Moreover, by carrying out polarization measurements in a second-generation experiment, one can test the assumption of V-A couplings, which thus does not have to be made blindly. Alternatively, if no additional peak is observed, to a given accuracy, then one can set a correlated upper bound on $|U_{ai}|^2$ for fixed $m(\nu_i)$. Because of the slow falloff of two-body phase space, there is little phase-space suppression of $M^+ \to \ell^+ \nu_i$ modes until $m(\nu_i)$ approaches the maximum allowed value. Indeed, for $\ell = e$, there is very drastic helicity enhancement of the rate

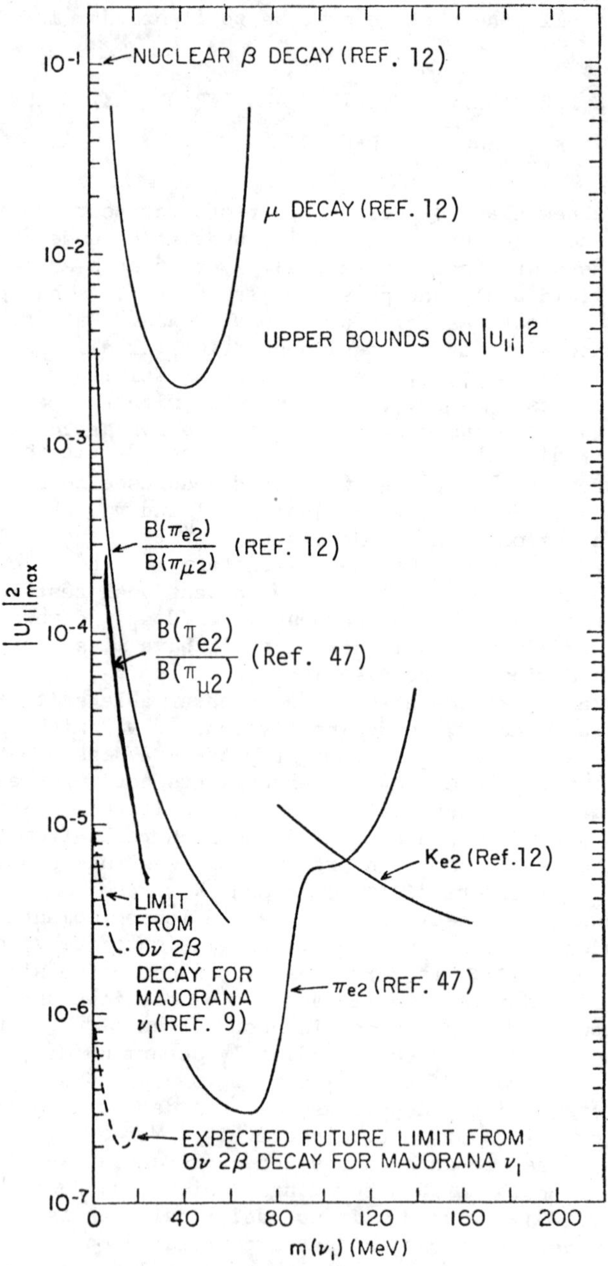
NUCLEAR β DECAY (REF. 12)
μ DECAY (REF. 12)
UPPER BOUNDS ON |U1i|²
B(πe2)/B(πμ2) (REF. 12)
B(πe2)/B(πμ2) (Ref. 47)
Ke2 (Ref.12)
LIMIT FROM 0ν 2β DECAY FOR MAJORANA ν1 (REF. 9)
πe2 (REF. 47)
EXPECTED FUTURE LIMIT FROM 0ν 2β DECAY FOR MAJORANA ν1
|U1i|²max
m(ν1) (MeV)
10-1
10-2
10-3
10-4
10-5
10-6
10-7
0
40
80
120
160
200

Figure 2

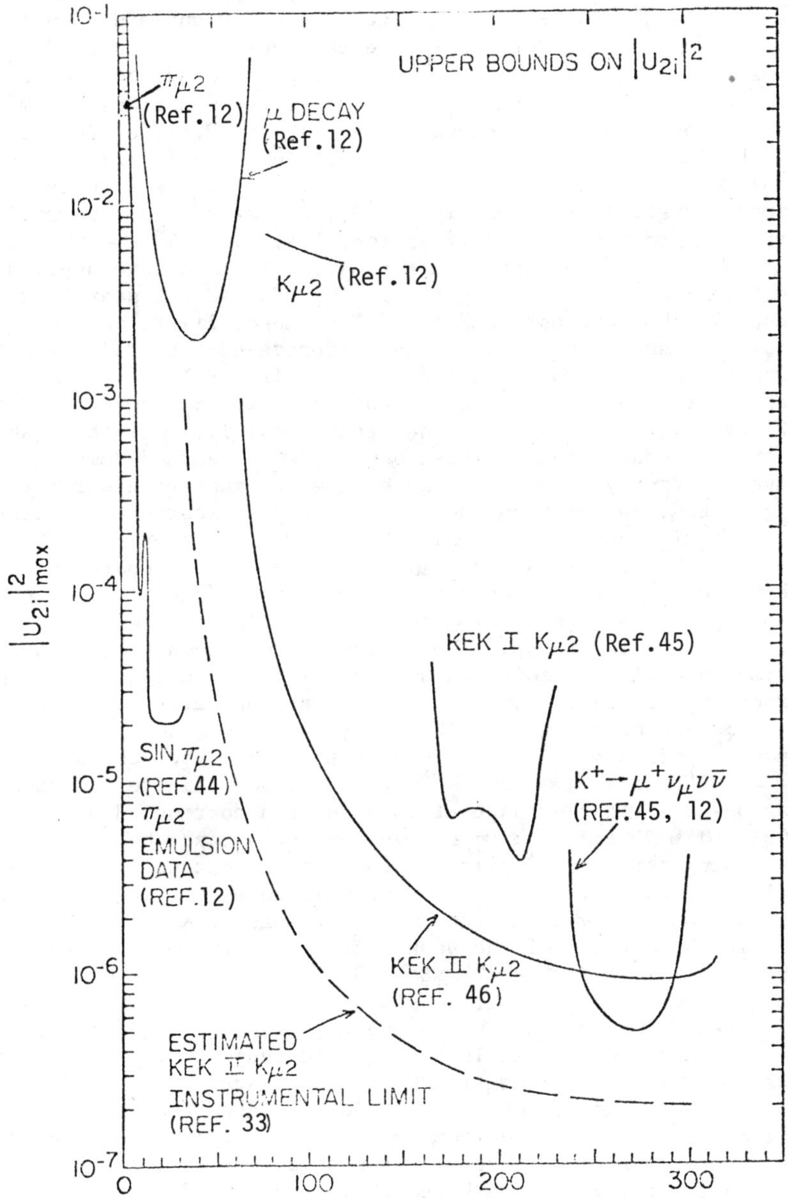
UPPER BOUNDS ON $|U_{2i}|^2$
$\pi_{\mu 2}$
(Ref.12)
μ DECAY
(Ref.12)
$K_{\mu 2}$ (Ref.12)
KEK I $K_{\mu 2}$ (Ref.45)
SIN $\pi_{\mu 2}$
(REF.44)
$\pi_{\mu 2}$
EMULSION
DATA
(REF.12)
$K^+ \to \mu^+ \nu_\mu \nu \bar{\nu}$
(REF.45, 12)
KEK II $K_{\mu 2}$
(REF. 46)
ESTIMATED
KEK II $K_{\mu 2}$
INSTRUMENTAL LIMIT
(REF. 33)
$|U_{2i}|^2_{max}$
$m(\nu_i)$ (MeV)
10^{-1}
10^{-2}
10^{-3}
10^{-4}
10^{-5}
10^{-6}
10^{-7}
0
100
200
300

Figure 3

factor for massive neutrino modes relative to that for massless neutrinos.

After the test was proposed, it was applied to existing data on $\pi_{\mu 2}$, $K_{\mu 2}$, and K_{e2} decays to search for any additional peaks within the cuts. (Data on π_{e2} decay could not be used for technical reasons; see Ref. 12). No such peaks were found, and correlated constraints on $|U_{1i}|^2$ and $|U_{2i}|^2$ were set as functions of $m(\nu_i)$. These upper bounds are shown in Figs. 2 and 3. The most stringent of the limits reached the 10^{-5} level for $|U_{2i}|^2$ (from $\pi_{\mu 2}$ emulsion data) and $\sim 3 \times 10^{-6}$ for $|U_{1i}|^2$ (from K_{e2} data). An additional constraint, primarily on $|U_{1i}|^2$, arose from the ratio of branching ratios $B(\pi_{e2})/B(\pi_{\mu 2})$ (see Fig. 2).

We proceed to discuss the experiments that have applied the peak search test. A group at SIN[44] performed the search in $\pi_{\mu 2}$ decay in 1981 and obtained the 90% CL upper limits on $|U_{2i}|^2$ for a heavy ν_i shown in Fig. 3. These improve upon the limits established from the analysis of data for 4 MeV $\lesssim m(\nu_i) \lesssim$ 12 MeV and are comparable to the limits which we obtained from $\pi_{\mu 2}$ emulsion data for 12 MeV $\lesssim m(\nu_i) \lesssim$ 34 MeV. The great sensitivity of the peak search test is demonstrated by these bounds, which extend down to the 10^{-5} level and apply for a decay mode in which massive neutrinos have so significant helicity enhancement. Two peak search experiments have been carried out by T. Yamazaki and collaborators at KEK on $K_{\mu 2}$ decay.[45,46] The first used an already existing apparatus and determined $|\vec{p}_\mu|$ via a range measurement technique.[45] It achieved a good upper limit on $|U_{2i}|^2$ in the range 160 MeV $\lesssim m(\nu_i) \lesssim$ 230 MeV, as indicated in Fig. 3. The Yamazaki group then proceeded to perform a beautiful, dedicated high-precision magnetic spectrometer experiment in early spring, 1982.[46] The apparatus for this experiment is shown in Fig. 1 of Ref. 46; there is a substantial quantity of NaI(Tl) crystals to veto events from the decays $K^+ \to \pi^0 \mu^+ \nu_\mu$; $\pi^0 \to \gamma\gamma$ and $K^+ \to \mu^+ \nu_\mu \gamma$. No definite additional peak was observed in the range of muon momenta corresponding to $m(\nu_i) \epsilon$ (60,320) MeV and an extremely good correlated upper bound was thus set on mixing strength $|U_{2i}|^2$ in this range, extending down to $\sim 10^{-6}$ (see Fig. 3). This group will take data again in 1983. It is anticipated that this further work at KEK will enable one to widen the range of the peak search down to $m(\nu_i) \sim$ 50 MeV and up to $m(\nu_i) \sim$ 400 MeV.[46] Beyond that, there are two possibilities for further work. First, consideration is being given to a TPC type of detector which would have greater efficiency for detecting photons and thereby vetoing background events. It is estimated that such a third-generation $K_{\mu 2}$ peak search experiment might be able to push the 90% CL limit on $|U_{2i}|^2$ down to the 10^{-7} level for $m(\nu_i)$ in the range from $\sim$ 200 to $\sim$ 300 MeV, and achieve commensurate improvements for lower masses. Second, there is perhaps some possibility of a peak search in the K_{e2} decay mode. Although a massive neutrino signal would be kinematically enhanced by a factor as large as $\sim 10^5$ relative to a massless neutrino mode, there is the challenge of a very small overall branching ratio with which one must contend.

A peak search has also been conducted recently by a group at TRIUMF using π_{e2} decay.[47] The upper limits on $|U_{1i}|^2$ from this experiment are shown in Fig. 2. The reason that the bound becomes much weaker for $m(\nu_i) \gtrsim 80$ MeV is that the background from the dominant decay chain $\pi^+ \to \mu^+ \to e^+$ sets in at a serious level in this lower e^+ momentum region, thereby reducing the sensitivity of the experiment to a small peak. Nevertheless, the power of the peak search method is again demonstrated by the extremely small limit of $|U_{1i}|^2 \lesssim 3 \times 10^{-7}$ which was established for 60 MeV $\lesssim m(\nu_i) \lesssim$ 75 MeV. In addition, a new measurement of $B(\pi_{e2})/B(\pi_{\mu 2})$ has been performed at TRIUMF which yields an improved upper limit on $|U_{1i}|^2$ for $m(\nu_i) \lesssim$ 40 MeV.

This, then, is the present state of experiments that have applied the peak search test for heavy neutrinos. In Europe there is a μ-capture experiment being performed by J. Deutsch and collaborators at SIN.[48] Since the final state is two-body, this group will again perform a search for anomalous peaks in the kinetic energy spectra of the recoiling nucleus. This experiment will directly probe the ν_i mass range below 100 MeV, and will be especially useful in the interval 34 MeV < $m(\nu_i) \lesssim$ 60 MeV, where only the previous analysis[12] of old $K_{\mu 2}$ data and μ decay data set upper limits on $|U_{2i}|^2$.

Concerning the future, as with other experiments searching for possible rare decays, peak searches in $K_{\ell 2}$ (and $\pi_{\ell 2}$) decays would profit from higher intensity beams. However, in order to utilize such higher intensities, they would need corresponding improvements in detectors, including, in particular, better photon detection and vetoing capabilities.

It is obvious that the flurry of excitement about nonzero neutrino masses which was generated by the reactor experiment of Reines and collaborators[49] has dissipated, now that their reported positive effect has been refuted by the Grenoble and Gösgen experiments.[50] The nonzero value of "$m(\nu_e)$" claimed by the ITEF experiment on ^{3}H beta decay also still awaits confirmation (or refutation) by independent experiments.[51] Nevertheless, it is certainly true that the issues of possible nonzero neutrino masses of Dirac or Majorana type, and associated lepton mixing, are quite general and will remain of continuing interest.

V SEARCHES FOR MASSES OF DOMINANTLY COUPLED NEUTRINOS IN $\pi_{\mu 2}$ AND $K_{\mu 3}$ DECAYS

The dominantly coupled neutrino mass eigenstate that can be studied usefully in K and π decays is ν_2, corresponding to the gauge group eigenstate ν_μ. Upper bounds on "$m(\nu_\mu)$", or more precisely, $m(\nu_2)$, have come historically from three sources, (and predominantly from the third): (1) measurement of the endpoint of the e^+ momentum spectrum in μ^+ decay[52]; (2) measurement of the $\pi\mu$ invariant mass distribution in $K^0_{\mu 3}$ decay;[53] and (3) measurement of $|\vec{p}_\mu|$ in $\pi_{\mu 2}$ decay.[54,55] The most recent and best such limit was obtained by Daum et al.[54] in a precision magnetic spectrometer experiment at SIN on $\pi_{\mu 2}$ decay and was reported as "$m(\nu_\mu)$" < 0.57 MeV (90% CL). This limit relied upon the value of m_{π^+} which was then available.

Subsequent to this experiment, a new measurement of m_{π^-} was carried out; if one combines this with previous measurments to obtain a new world average and then uses that quantity in conjunction with the SIN data from Ref. 54, one obtains the slightly lower limit, $m(\nu_2) < 0.52$ MeV (90% CL).[55]

A new $\pi_{\mu 2}$ experiment is now under development by P. Nemethy and collaborators at LBL.[56] It will utilize a ring-imaging Čerenkov counter and anticipates achieving an ultimate sensitivity of 50 KeV in $m(\nu_2)$. Thus, in the near future, one may look forward to a dramatic reduction in the upper limit on $m(\nu_2)$ (or, of course, possibly an observation of a nonzero value of this mass).

In addition to $\pi_{\mu 2}$ decays, one might consider the possibility of using $K_{\mu 3}$ decays to get an improved limit on $m(\nu_2)$. The last such limit was obtained in 1974 by an LBL group using $K^o_{\mu 3}$ decay and is "$m(\nu_\mu)$" < 1.5 MeV (90% CL). Some features of such an experiment were reviewed at the DPF Snowmass Summer Study.[9,57] However, assuming that the LBL-BNL experiment of Nemethy et al. does achieve a level of sensitivity roughly equal to its expectations, it is not clear that a $K_{\mu 3}$ experiment would be competitive. Thus, summarizing, it would be of interest to consider extensions of the dedicated $\pi_{\mu 2}$ experiment of Ref. 56 which might be designed for a pion port at a high luminosity kaon factory or kaon beam at an upgraded AGS.

VI OTHER BOUNDS ON NEUTRINO MASSES AND MIXING FROM MUON AND NEUTRINO DECAYS

In order to assess possible future limits on neutrino masses and mixing which might be obtained from K and π decays, one must also take account of bounds from other sources. One such bound was derived in 1980-1981 from an analysis of muon decay data.[12] The basic starting point of this analysis was the realization that in general "the" decay $\mu^- \rightarrow \nu_\mu e \bar{\nu}_e$ really consists of a subset of all of the modes $\mu^- \rightarrow \nu_i e \bar{\nu}_j$ which are allowed by phase space, where ν_i, etc. are the neutrino mass eigenstates. Now, if one or more neutrinos with non-negligible masses is emitted, it will alter the overall, observed distribution in a calculable manner. Hence, the data analyses that were carried out to test for anomalous Lorentz structure in leptonic weak interactions would have derived apparent non-(V-A) values for the various spectral parameters in muon decay, given the fact that they fit the data to the well-known zero neutrino mass distributions depending on ρ, η, ξ, and δ. It was thus possible to use the agreement between the measured values of these spectral parameters and the V-A values (with radiative corrections taken into account to the requisite degree of accuracy) to place upper limits on any admixture, via lepton mixing, of heavy neutrino decay modes. It was found that the parameter ρ was the most sensitive to massive neutrino effects, and this was utilized to set an upper bound on $|U_{ri}|^2$, r = 1 and 2, for $m(\nu_i)$ in the range 10 MeV < $m(\nu_i)$ < 70 MeV. This bound is shown in Figs. 2 and 3 for the respective cases r = 1 and r = 2. For the latter case, i.e., the coupling of a heavy neutrino to the muon, it provides the primary constraint on this coupling and the associated lepton mixing in the region of $m(\nu_i)$ covered. In contrast to the peak search experiments discussed above, which do not directly probe the charge conjugation properties of the neutrinos, muon decay does. A general analysis of the decay distribution for the case of massive neutrinos of both Dirac and Majorana types, lepton mixing, and arbitrary Lorentz structure has been given.[58] The distribution in the case of Majorana neutrinos and V ± A couplings was also studied by the Osaka group of Doi et al.[59] Finally, a special case of the decay into Dirac neutrinos analyzed in Ref. 12 was later considered by Kalyniak and Ng, who also attempted to calculate the decay distribution for the Majorana case with V-A couplings.[60] However, the results of Kalyniak and Ng are incorrect, owing to their failure to take proper account of the self-conjugacy of Majorana neutrinos, as was noted in Ref. 58.

There are currently several very high-precision experiments on regular muon decay at TRIUMF, LAMPF, and SIN. These will be discussed further in Section VIII below. The increased accuracy which these experiments will achieve in the measurement of the muon decay distribution, and thus the associated spectral parameters (especially ρ) will make possible a commensurate improvement of the bounds on leptonic mixing angles via the method of Ref. 12. These experiments will be especially valuable since, together with the ongoing SIN muon-capture experiment of J. Deutsch and collaborators, they will constrain $|U_{2i}|$ over a range of $m(\nu_i)$ between $\sim$ 34 MeV and $\sim$ 60 MeV not covered at all in $\pi_{\mu 2}$ and below the optimally sensitive region of neutrino masses covered in $K_{\mu 2}$ peak search experiments. The TRIUMF experiment of M. Strovink and collaborators[61] is now reporting results; the others should have results within a few years. As will be stressed in Section VIII, tests of lepton mixing are but one of many reasons for guaranteeing a high intensity muon beam as an auxiliary part of a future kaon factory.

The future prospects for further neutrino oscillation experiments have been discussed in detail in the LANL <u>Proposal for a National Facility to Provide a High Intensity Neutrino Source</u>[62], by R. Lanou in the DPF Snowmass Summer Study[63], and for this workshop by the Kayser-Rosen subgroup.[64] Neutrino oscillation experiments nicely complement searches for effects of neutrino masses and lepton mixing in particle decays since they are sensitive to much smaller masses (more precisely, mass differences), but cannot yet probe for such small mixing angles as peak search experiments. An experiment to look for the decays of massive neutrinos is currently under consideration by the CERN PS. The revised version[65] of the proposal for this experiment envisions a search for the decay $\nu_i \to \nu_j e^+ e^-$; the ν_i thus cannot be the primary mass eigenstate in either ν_e or ν_μ and hence would be produced only subdominantly in π or K decays.[12] Clearly, given the upper limits that have been established on lepton mixing for such massive neutrinos, the production of the initial ν_i's would be very strongly suppressed. Since the decay $\nu_i \to \nu_j e^+ e^-$ itself only only proceeds via lepton mixing, it is also suppressed by a lepton mixing matrix coefficient squared (for the dominant mode, in which j = 1). However, assuming this experiment is approved to run, it will be interesting to see what results it obtains.

VII OTHER K DECAYS, AND SOME π DECAYS, OF INTEREST

In this section we shall mention some further K decays which might be of interest for a dedicated kaon facility. Although the primary purpose of this workshop is to assess the physics uses of a high-intensity source of strangeness, it does seem worthwhile to list several pion decays which could be searched for or studied via an auxiliary pion beam at such a facility. Obviously, experimental programs on pion and muon decays do not require K beams, but if one is envisioning a high-intensity, low-energy machine, it is important to retain a strong continuing effort in these two other areas. We proceed with several conventional kaon decays:

(1) $K^+ \to e^+ \nu_e \gamma$

This radiative decay is useful for studying structure-dependent electromagnetic corrections to two-body leptonic K decays. Such information is valuable in order to make a precise comparison between the V-A predictions and experimental measurements for the ratio $B(K_{e2})/B(K_{\mu 2})$. Since structure-dependent corrections are expected to be substantially larger for this ratio than for the corresponding pion ratio $B(\pi_{e2})/B(\pi_{\mu 2})$, and since they are hard to calculate reliably, further experimental input is helpful. The $K^+ \to e^+ \nu_e \gamma$ decay was studied recently in a CERN-Heidelberg experiment[67]; the current value of the branching ratio for the structure-dependent contribution involving photons of positive helicity is[55] $(1.52 \pm 0.23) \times 10^{-5}$

(2) $K_L^o \to \gamma\gamma$; $K^+ \to \pi^+ \gamma\gamma$

In the quasi-free quark model approach, the decay $K_L^o \to \gamma\gamma$ proceeds via one-loop electroweak diagrams. It has the interesting feature that the GIM suppression mechanism operates differently than in the case of $K^o - \bar{K}^o$ mixing or the decays $K_L^o \to \mu^+\mu^-$ and $K^+ \to \pi^+ \nu\bar{\nu}$. Rather than a multiplicative suppression factor $\propto (m_c^2 - m_u^2)/m_W^2$ in the amplitude (for two generations), Gaillard and Lee found a different dependence, involving m_q^2/m_K^2 but not additional m_W^{-2} suppression.[16] This absence of a severe GIM suppression was seen to be in agreement with the measured value of the branching ratio for this decay, the present value of which[55] is $B(K_L^o \to \gamma\gamma) = (4.9 \pm 0.4) \times 10^{-4}$. A similar one-loop electroweak K decay is $K^+ \to \pi^+ \gamma\gamma$, for which Gaillard and Lee estimated a branching ratio, in the two-generation case, $\sim 10^{-6}$-10^{-7}. As with other rare K decays, the free-quark contribution to this decay could be somewhat larger for the present three-generation theory due to the t-quark terms. The current upper limit, established recently by a KEK experiment,[29] is $B(K^+ \to \pi^+\gamma\gamma) < 0.84 \times 10^{-5}$ (90% CL). It would certainly be worthwhile to perform an experiment with improved sensitivity in order actually to observe this decay.

(3) $K_L^o \to K^{\pm} e^{\mp} \overset{(-)}{\nu}_e$

The branching ratio for kaon beta decay is calculated to be extremely small: $\sim 3 \times 10^{-9}$. The potential value of an experimental measurement of the branching ratio for this decay is that the latter depends quite sensitively on the mass difference between the neutral and charged kaons, as $(m_{K^o} - m_{K^+})^5$, approximately. Hence, a sufficiently precise measurement of the decay branching ratio could provide a more accurate determination of this mass difference than the one now available, viz., $m_{K^o} - m_{K^+} = 4.01 \pm 0.13$ MeV.

Rare K decays which yield information concerning CP violation are discussed in detail in the corresponding section of these Proceedings, chaired by L. Wolfenstein. Accordingly, we shall not analyze such decays here. We proceed to consider briefly some interesting pion decays which might deserve further study at a pion port in a future high-intensity, low-energy facility:

(4) $\pi^+ \to \pi^o e^+ \nu_e$

Pion beta decay has provided one of the important tests of the conserved vector current hypothesis. The current value of the branching ratio for this decay is $1.02 \pm 0.07 \times 10^{-8}$. It would clearly be worthwhile to improve the precision with which this number can be measured, to carry further the comparison with theory.

(5) $\pi^o \to e^+ e^-$

Until recently, the only observation of this rare decay came from a CERN-Geneva-Saclay experiment.[68] A LAMPF experiment has now reported[69] a branching ratio $B(\pi^o \to e^+e^-) = (1.8 \pm 0.6) \times 10^{-7}$. Although the rate for this decay cannot be calculated with quite the precision of pion beta decay, more data would certainly be helpful in understanding the mechanisms responsible for it.

(6) $\pi^o \to \gamma\gamma\gamma$

This decay mode provides a test of charge conjugation invariance in the electromagnetic interactions. The present upper limit on the branching ratio is 3.8×10^{-7}, established by a recent LAMPF experiment.[70] Because of the additional factor of α (relative to the rate for the regular decay $\pi^o \to \gamma\gamma$), the suppression due to three-body phase space, and other factors, the branching ratio would be expected to be very small even if charge conjugation invariance were violated in electromagnetic interactions. Therefore, it is worthwhile to push this upper bound down further in dedicated future experiments.

(7) $\pi^o \to$ missing neutrals

A specific exmaple of this decay would be $\pi^o \to \nu\bar{\nu}$, either involving anomalous Lorentz structure[71] or massive neutrinos.[72] The obvious problems are (1) how to tag the π^o with sufficiently high reliability, and (2) how to detect and veto the regular $\gamma\gamma$ decay with the requisite extrem high efficiency. The decay $K^\pm \to \pi^+\pi^o$ might be used to provide the taggir

VIII MUON DECAYS AND REACTIONS

Regular and rare muon decays serve as probes for much of the same new physics that rare K decays do. We shall briefly mention some muon decays of interest in this section:

(1) "Regular" Muon Decay: $\mu^+ \to e^+$ + missing neutrals

The conventional source for the experimentally observed final state in this decay is $\mu^+ \to \bar{\nu}_\mu e^+\nu_e$, where the neutrinos are massless, and the weak couplings are of V-A type. Exotic possibilities include possible right-handed current contributions, or more generally, effective interactions with anomalous Lorentz structure, and effects of massive neutrino and associated lepton mixing. Indeed, if the latter are present, then "the" decay $\mu^+ \to \nu_\mu e^+\nu_e$ is not one decay at all, but rather a sum of all the separate modes $\mu^+ \to \bar{\nu}_i e^+ \nu_j$, where the ν_i's are neutrino mass eigenstates. The observed e^+ momentum spectrum is thus the sum due to all the actual (i,j) modes and thus differs from the spectrum predicted by the standard model. Consequently, the values of the spectral parameters ρ, η, ξ, and δ which were inferred from the data by fitting it to an assumed standard-model form with generalized Lorentz structure would differ from the V-A values even if the couplings were purely of V-A type! These effects were pointed out in Ref. 12 and were used to derive correlated bounds on neutrino masses and lepton mixing, as was discussed in Section VI. The upper bounds on $|U_{ri}|^2$, r = 1 and 2, which were obtained are shown in Figs. 2 and 3, as functions of $m(\nu_i)$. Moreover, in the context of supersymmetric theories, the decay $\mu^+ \to e^+\tilde{\gamma}\tilde{\gamma}$ could occur, if the photinos are light enough, and would yield a generic final state of the form e^+ + missing neutrals if neither of the photinos decayed in the detector. The momentum spectrum of the positrons would again differ from the standard model prediction, both because of generally different effective Lorentz structure and because of possible nonzero photino masses.

There are currently three new experiments on regular muon decay, by M. Strovink and collaborators at TRIUMF,[61], K. Crowe and collaborators, also at TRIUMF,[73] and H. Anderson and collaborators, at LAMPF.[74] The first of these has obtained the result $\xi P_\mu \delta/\rho > 0.9959$ (90% CL).[61] The second will measure η, while the LAMPF experiment plans to measure each of the four spectral parameters. One should also mention the recent measurement of the longitudinal polarization of the positron in muon decay at SIN, and the fact that this group plans to carry out an experiment measuring the spectral parameters in the near future.[75] The new measurement by Strovink et al. provides a sensitive limit on possible contributions due to right-handed currents. The results from the other experiments listed should be forthcoming in the near future and will provide similarly useful constraints on possible new physics.

(2) $\mu^+ \to e^+\gamma$, $\mu^+ \to e^+e^+e^-$, and $\mu^+ \to e^+\gamma\gamma$

These decays are all forbidden by lepton family number conservation. Many of the new physics possibilities which would give rise to rare K decays such as $K^+ \to \pi^+\mu^+e^-$, $K^0_L \to \mu^+e^-$, etc., also would cause these decays to occur at some level. The decays $\mu^+ \to e^+\gamma$ and $\mu^+ \to e^+e^+e^-$ were analyzed in the standard $SU(2)_L \times U(1)$ electroweak theory, extended to include massive neutrinos and lepton mixing, in Refs. 11 and 76; and in theories with right-handed currents in Ref. 77. A discussion of how these decays proceed in theories with horizontal gauge interactions was given in Ref. 8. Effects expected in (extended) technicolor models have been considered in Ref. 15.

The current upper limit $B(\mu^+ \to e^+\gamma) < 1.9 \times 10^{-10}$ was achieved by a LAMPF experiment,[78] while the bound $B(\mu^+ \to e^+e^+e^-) < 1.9 \times 10^{-10}$ was reported by a Dubna group.[79] The present upper limit on the two-photon mode is[80] $B(\mu^+ \to e^+\gamma\gamma) < 5 \times 10^{-8}$. There is now a dedicated experiment running at LAMPF searching for each of these three decays.[81] It anticipates a sensitivity in the 10^{-11}-10^{-12} range for the branching ratios in all three decays. Results from this experiment should be available on a time scale of 1-2 years. It is further anticipated that with the requisite improvements in detectors, an experiment using a muon beam at a high-intensity, low-energy facility in the early 1990's might be able to achieve a sensitivity of 10^{-13} or better in branching ratio.[82]

(3) $\mu^- N \to e^{\pm} N'$

Like the muon decays discussed above in category (2), muon conversion in the field of a nucleus (N) violates lepton family number. The second reaction, yielding an e^+, also violates total lepton number. The first reaction was analyzed in the extended standard model in Ref. 83; other sources for both reactions have been discussed in several of the works cited above. The current upper limit $\sigma(\mu^- + {}^{32}S \to e^- + {}^{32}S)/\sigma(\mu^- + {}^{32}S \to \nu_\mu + {}^{32}P) < 7 \times 10^{-11}$ has been achieved in an experiment at SIN.[84] A current experiment at TRIUMF anticipates reducing this limit to[85] 10^{-12}. It is further estimated that the limit might be reduced to 10^{-13} or better at the high-intensity facility under consideration here, but only if one could achieve sufficient improvements in the necessary detectors.[85,86]

Thus, in summary, the study of regular muon decay and the search for possible rare decays are worthwhile activities to pursue at a future high-intensity facility. We would like to thank T.P. Cheng and L.F. Li for helpful input for this section.

IX. CONCLUSION

The overall conclusions are summarized in Table 1 presented earlier. In ending this report, we would like to stress that experimentalists should search for as wide a range of new physics as possible and should not be deterred from considering a particular decay mode just because some theorist claims that it is predicted not to occur in his set of fashionable models. New physics has sometimes been anticipated by theorists, but at least as often, it has not. Moreover, theorists may not correctly assess the experimental implications of a given model. It is instructive to recall, for example, that $|\varepsilon'/\varepsilon|$ was once claimed to be negligibly small in the standard electroweak theory,[86] so that there was no reason, supposedly, for experimentalists to search for nonzero ε'. This claim was shown to be incorrect by Gilman and Wise,[87] and subsequently we have seen a great resurgence of interest in CP-violation experiments, as was discussed at this workshop by Winstein and Wolfenstein. The moral of this story should be clear. It is also clear that, despite some theorists' possible negative views, it is worthwhile to search for new physics via rare K (and π and μ) decays to as low levels as allowed by beam intensities, backgrounds, and detector capabilities. An upgraded AGS and dedicated kaon factory would constitute very powerful research tools for this purpose.

REFERENCES

The following references are not to be taken as an exhaustive compilation. They are only intended as an initial guide to the theoretical literature, especially for those who are unfamiliar with certain topics.

1. Some reviews which may be of help to the reader are listed below: Elementary Particle Physics and Future Facilities, Proceedings of the APS Division of Particles and Fields Workshop, Snowmass, June-July, 1982 R. Donaldson, R. Gustafson, and F. Paige, eds. (in particular, the discussions of standard model predictions and tests of theoretical developments beyond the standard model); Proceedings of the Second LAMPF II Workshop, H. A. Thiessen, et al., eds. Los Alamos Pub. LA-9572-C. LAMPF II Proposal, LANL preprint. For a review of technicolor, see, e.g., E. Farhi and L. Susskind, Phys. Rep. 74C, 277 (1981). For supersymmetry, see, for example, the discussion of theoretical rudiments in P. Fayet and S. Ferrara, Phys. Rep. 32C, 249 (1977) and, for supergravity, P. van Nieuwenhuizen, Phys. Rep., 68C, 189 (1981). For up-to-date reviews of the implications for particle physics, see, for example, G. Kane and J. Polchinski in the Proceedings of the Fourth Workshop on Grand Unification, to be published.
2. The possibility of such a mass relation (based on using just one or more 5's of Higgs, but no 45's) was first noticed in the classic SU(5) paper of H. Georgi and S. Glashow, Phys. Rev. Lett. 32, 438 (1974) and later rejected in H. Georgi and C. Jarlskog, Phys. Lett. 86B, 297 (1979). The statement that one should use only a 5 of Higgs and the claim that the generic mass relation is correct is due to A. Buras, J. Ellis, M. Gaillard, and D. Nanopoulos, Nucl. Phys. B135, 66 (1978).
3. See e.g. C. Burges and H. Schnitzer, Brandeis preprint (1983); and E. Eichten, K. Lane, and M. Peskin, Phys. Rev. Lett. 50, 811 (1983); and references therein.
4. A. R. Clark et al. Phys. Rev. Lett. 26, 1667 (1971).
5. M. E. Zeller, spokesman, "Search for the Rare Decay Mode $K^+ \to \pi^+\mu^+e^-$", BNL AGS Experiment E777.
6. W. Morse and M. P. Schmidt, co-spokesmen, "A Search for the Flavor-Changing Neutral Currents $K_L^o \to \mu$ +e and $K_L^o \to e^+e^-$.
7. M. Schmidt, private communication.
8. R. Cahn and H. Harari, Nucl. Phys. B176, 135 (1980). Similar limits are given in G. Kane and R. Thun, Phys. Lett. 94B, 513 (1980) and D. R. T. Jones, G. Kane, and J. Leveille, Nucl. Phys. B198, 45 (1982). See also P. Herczeg, in Proceedings of the Kaon Factory Workshop, TRIUMF (1979), M. Craddock, ed., and O. Shanker, Nucl. Phys. B185, 382 (1981). A number of the limits given by Shanker involve unconventional assumptions and calculations.
9. For a review, see R. Shrock, in the Proceedings of the DPF Summer Study on Elementary Particle Physics and Future Facilities,

eds., R. Donaldson, R. Gustafson, and F. Paige, June-July, 1982, p. 264; and Ref. 10.

10. T. Yamazaki, in the Proceedings of the International Conference on Neutrino Physics and Astrophysics, 1982.
11. B. W. Lee and R. Shrock, Phys. Rev. D16, 1444 (1977).
12. R. Shrock, Phys. Lett. 96B, 159 (1980); Phys. Rev. D24, 1232, 1275 (1981).
13. E. Eichten and K. Lane, Phys. Lett. 90B, 125 (1980).
14. S. Dimopoulos, G. Kane, and S. Raby, Nucl. Phys. B182, 77 (1981).
15. J. Ellis et al., Phys. Lett. 101B, 387 (1981); S. Dimopoulos and J. Ellis, Nucl. Phys. B182, 505 (1981).
16. M. K. Gaillard and B. W. Lee, Phys. Rev. D10, 897 (1974).
17. B. W. Lee, J. Primack, and S. B. Treiman, Phys. Rev. D7, 510 (1973) (which considered rare induced K decays in the old Georgi-Glashow O(3) model); A. Vainstein and I. Khriplovich, Pis'ma Zh. Eksp. Teor. Fiz. 18, 41 (1973) (JETP Lett. 18, 83 (1973)).
18. R. Dzhelyadin et al., Phys. Lett., 105B, 239 (1981).
19. For early reviews, see M. K. Gaillard and H. Stern, Ann. Phys. 76, 580 (1973); A. Dolgov, V. Zakharov, and L. Okun, Usp. Fiz. Nauk 107, 537 (Sov. Phys. Usp. 15 404 (1973)).
20. R. Shrock and M. Voloshin, Phys. Lett., 87B, 375 (1979); T. Inami and C. S. Lim, Prog. Theor. Phys. 65, 297 (1981) Concerning the long-distance contribution to Re $\{Amp(K_L^o \to \mu\bar{\mu})\}$, see M. B. Voloshin and E. P. Shabalin, Pis'ma Zh. Eksp. Teor. Fiz. 23 (1976) 123 [JETP Lett. 23 (1976) 107].
21. R. Carithers et al., Phys. Rev. Lett. 31, 1025 (1973)
22. Y. Fukashima et al., Phys. Rev. Lett. 36, 348 (1976).
23. M. Shochet et al., Phys. Rev. Lett. 39, 59 (1977); Phys. Rev. D19, 1965 (1979).
24. C. Quigg and J. D. Jackson, UCRL Report 18487 (1968); L. Sehgal, Phys. Rev. 183, 1511 (1969); B. Martin, E. deRafael, and J. Smith, Phys. Rev. D2, 179 (1970); M. K. Gaillard, Phys. Lett. 35B, 431 (1971); G. Farrar and S. B. Treiman, Phys. Rev. D4, 257 (1971); M. Pratap, J. Smith, and Z. Uy, Phys. Rev. D5, 269 (1972); S. Adler, G. Farrar, and S. B. Treiman, Phys. Rev. D5, 770 (1972); L. Bergstrom, Zeit. Phys. C14, 129 (1982).
25. M. K. Gaillard, B. W. Lee, and R. Shrock, Phys. Rev. D13, 2674 (1976).
26. P. Herczeg, Phys. Rev. D27, 1512 (1983).
27. R. Peccei and H. Quinn, Phys. Rev. Lett. 38, 1440 (1977); Phys. Rev. D16, 1791 (1977); S. Weinberg, Phys. Rev. Lett. 40, 223 (1978); F. Wilczek, ibid., 40, 279 (1978).
28. F. Wilczek, Phys. Rev. Lett. 49, 1549 (1982).
29. Y. Asano et al., Phys. Lett. 107B, 159 (1981).
30. S. Weinberg, op cit., Ref. 27; T. Goldman and C.M. Hoffman, Phys. Rev. Lett., 40, 220 (1978); W. Bardeen and S.-H. H. Tye, Phys. Lett. 74B, 229 (1978); J. Kandaswamy et al., Phys. Lett. 74B , 377 (1978); J. Frere et al., Phys. Lett. 103B, 129 (1981).
31. The KEK experiment of Ref. 29 obtained the limits $B(K^+ \to \pi^+\gamma\gamma) < 0.84 \times 10^{-5}$ and $B(K^+ \to \pi^+a;\ a \to n\gamma's) < 1.4 \times 10^{-6}$ (both at the

90% CL) for $m_a < 100$ MeV and $\tau_{a \to n\gamma's} < 10^{-9}$ sec. See also Ref. 33.

32. J. Kim, Phys. Rev. Lett. 43, 103 (1979); M. Dine et al., Phys. Lett. 104B, 199 (1981); H.P. Nilles and S. Raby, Nucl. Phys. B198, 102 (1982); M.B. Wise et al., Phys. Rev. Lett. 47, 402 (1981).
33. T. Yamazaki et al., University of Tokyo preprint UTMSL-35, and T. Yamazaki, private communication.
34. R. Shrock and M. Suzuki, Phys. Lett. 110B, 250 (1982).
35. E. Ma and J. Okada, Phys. Rev. D18, 4219 (1978).
36. R. Shrock, in the Proceedings of the DPF Summer Study on Elementary Particle Physics and Future Facilities, Snowmass, June-July, 1982, R. Donaldson et al., eds., p. 291.
37. J. Ellis and J. Hagelin, Nucl. Phys. B217, 189 (1983).
38. This negative conclusion was reached in the first work, that of Ma and Okada, which investigated the possibility of counting neutrinos via this decay. It was also reached in Ref. 36 upon a reexamination of the issue.
39. M. Suzuki, UC Berkeley-LBL preprint UCB-PTH-82/8.
40. M. K. Gaillard, Y.-C. Kao, I.-H. Lee, and M. Suzuki, Phys. Lett. 123B, 241 (1983).
41. G. Farrar and P. Fayet, Phys. Lett. 105B, 155 (1981) and references therein.
42. H. Goldberg, Phys. Rev. Lett. 50, 1419 (1983).
43. An up-to-date review of mass limits pertaining to supersymmetric partners of known particles is given in G. Kane, in the Proceedings of the Fourth Workshop on Grand Unification, 1983, to be published (op. cit., Ref. 1).
44. R. Abela et al., Phys. Lett. 105B, 263 (1981).
45. Y. Asano et al., Phys. Lett. 104B, 84 (1981).
46. R. Hayano et al., Phys. Rev. Lett. 82, 1305 (1982); T. Yamazaki, private communication.
47. D. Bryman et al., Phys. Rev. Lett. 50, 7 (1983); and TRIUMF preprint.
48. J. Deutsch, private communication and to be published.
49. F. Reines, H. Sobel, and E. Pasierb, Phys. Rev. Lett. 45, 1307 (1980).
50. J. Vuilleumier et al., Phys. Lett. 114B, 298 (1982); and to be published.
51. A review of ongoing ^{3}H beta decay experiments is given in Ref.9.
52. M. Bardon et al., Phys. Rev. Lett. 14, 449 (1965).
53. A. Clark et al., Phys. Rev. D9, 533 (1979).
54. M. Daum et al., Phys. Lett. 74B, 126 (1978); Phys. Rev. D20, 2692 (1979).
55. Particle Data Group, Review of Particle Properties, Phys. Lett. 111B, 1 (1982).
56. P. Nemethy, spokesman, letter of intent submitted to Brookhaven National Laboratory (September, 1982), and private communication.

57. C. Hoffman and V. Sandberg, in Proceedings of the 1982 DPF Summer Study on Elementary Particle Physics and Future Facilities, Snowmass, eds., R. Donaldson, R. Gustafson, and F. Paige, p. 552.
58. R. Shrock, Phys. Lett. 112B, 382 (1982).
59. M. Doi et al., Prog. Theor. Phys. 67, 281 (1982).
60. P. Kalyniak and J. Ng, Phys. Rev. D25, 1305 (1982).
61. J. Carr et al., LBL preprint in preparation, and M. Strovink, private communication.
62. LANL Working Group, G. J. Stephenson, chairman, Proposal to the Department of Energy for a National Facility to Provide a High Intensity Neutrino Source (Los Alamos National Laboratory, 1982).
63. R. Lanou, in the Proceedings of the 1982 DPF Summer Study, op. cit., p. 538.
64. B. Kayser and S. P. Rosen, in these Proceedings.
65. F. Vannuci, spokesman, CERN Experimental Proposal (1983), revised version.
66. T. Goldman and W. J. Wilson, Phys. Rev. D15, 709 (1977).
67. J. Heintze et al., Nucl. Phys. B149, 365 (1979).
68. J. Fischer et al., Phys. Lett. 73B, 364 (1978).
69. R. E. Mischke et al., Phys. Rev. Lett. 48, 1153 (1982).
70. V. L. Highland et al., Phys. Rev. Lett. 44, 628 (1980).
71. E. Fischbach, B. Kayser, G. Garvey, and S. P. Rosen, Phys. Lett. 52B, 385 (1974).
72. P. Herczeg and C. M. Hoffman, Phys. Lett. 100B, 347 (1981).
73. K. Crowe et al., TRIUMF Experimental Proposal.
74. H. Anderson et al., LAMPF Experimental Proposal.
75. F. Corriveau et al., Phys. Rev. D24, 2004 (1981); and F. Scheck, private communication. (For a review of previous experimental and theoretical results on muon decay, see F. Scheck, Phys. Rep. 44, 187 (1978).)
76. W. J. Marciano and A. I. Sanda, Phys. Lett. 67B, 303 (1977).
77. T. P. Cheng and L. F. Li, Phys. Rev. D16, 1425 (1977); J.D. Bjorken, K. Lane, and S. Weinberg, ibid., D16, 1474 (1977); S. B. Treiman, F. Wilczek, and A. Zee, ibid., D16, 152 (1977).
78. J. Bowman et al., Phys. Rev. Lett. 42, 556 (1979).
79. M. Korenchenko et al., JETP 43, 1 (1976).
80. J. Bowman et al., Phys. Rev. Lett. 41, 442 (1978).
81. J. Bowman et al., LAMPF Experiment.
82. G. Sanders and C. M. Hoffman, private communications.
83. W. J. Marciano and A. I. Sanda, Phys. Rev. Lett. 38, 1512 (1977).
84. A. Badertscher et al., Lett. Nuovo Cim. 28, 401 (1980).
85. D. Bryman et al., TRIUMF Experiment; D. Bryman, private communication.
86. J. Ellis, M.K. Gaillard, and D.V. Nanopoulos, Nucl. Phys. B109, 213 (1976).
87. F. Gilman and M. Wise, Phys. Lett. 83B, 83 (1979).

$K \to \pi$ + LIGHT PSEUDOSCALAR

J.-M. Frere
Randall Laboratory of Physics
University of Michigan, Ann Arbor, MI 48109

The "classical" or "standard" axion now looks very dead and I feel no urge to try and revive it. I will however comment briefly on the present experimental situation,[1] and envisage the sensitivity of the more general decay $K \to \pi$ + light pseudoscalar, since such light particles might appear in a considerable number of models or theories (e.g.: supersymmetry, technicolor ...).[2]

For this purpose, I will consider two cases:

i) the tree-level coupling of the pseudoscalar h is flavor-diagonal (axion-type)

ii) the tree-level coupling is not flavor diagonal (flavor changing neutral current).

FLAVOR DIAGONAL TREE LEVEL COUPLING

The relevant part of the Lagrangian reads:

$$\frac{h}{2m_W} \Sigma \{m_i \; x \; \bar{u}_i\gamma_5 u_i + m_j \; y \; \bar{d}_j\gamma_5 d_j\} \tag{1}$$

where

$$u_i = (u,c,t) \qquad d_j = (d,s,b)$$

The "standard" axion is described by: $|y| = |\frac{1}{x}|$; if more doublets are present, one usually has $|xy| > 1$. $SU(2) \times U(1)$ singlets with vacuum expectation values would allow $|xy| < 1$, but astrophysical data then imply $|xy| < 10^{-14}$, making earth-bound experiments somewhat hopeless.

The simplest and clearest way to see that the standard axion is now excluded is to consider the two branching ratios $B(\Upsilon \to h\gamma)$ and $B(\psi \to h\gamma)$. These tests are relatively clean because only the direct axion coupling to heavy quarks is involved, and the theoretical expectation is:

$$B(\Upsilon \to h\gamma)\; B(\psi \to h\gamma) \simeq (xy)^2 \frac{G_F^2 \; m_c^2 \; m_b^2}{2\pi^2 \; \alpha^2} B_{\mu\mu}(\psi) \cdot B_{\mu\mu}(\Upsilon) \simeq (xy)^2 \times 1.6 \times 10^{-8}$$

experimentally:[3]

$$\text{l.h.s.} < 0.9 \times 10^{-9} \text{ (90\% C.L.)} \tag{2}$$

and this excludes the "standard" axion for "all values of x". Note the quotation marks around "all values of x". Indeed some care should be taken in view of the possibly non-neglibible mass of the axion for

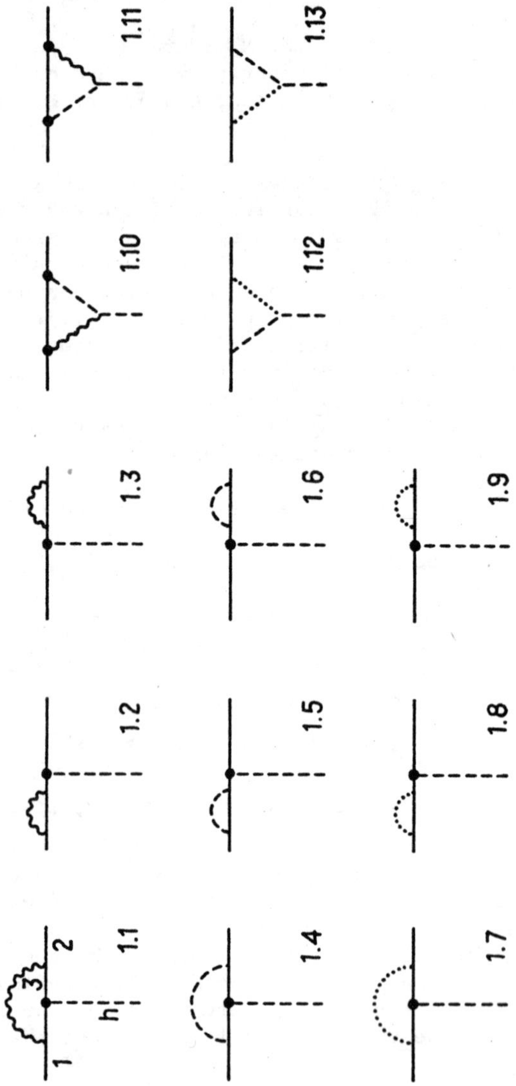

FIGURE 1

One-loop corrections to coupling of pseudoscalar (h) to quarks. (See Ref. 4.) The notation is: solid lines -- quarks; wavy lines -- gauge bosons; dashed lines -- pseudoscalar (h); and dotted lines -- charged scalar bosons ($H^{\pm}$).

x (or y) $\to \infty$, since this mass obeys (for N families of quarks and leptons):

$$m_h \simeq N(x+y) \times 23 \text{ keV}. \tag{3}$$

For large enough x or y, axions could then not be produced in ψ or Υ decays. It is however an easy matter to check that such values of x or y ($x \sim 10^5$ for $m \sim 1$ GeV) would not be compatible to a perturbative treatment of (1), therefore invalidating the whole approach.

If we want to depart from the relation xy = 1, with other possible models or light pseudoscalars in mind, we may ask the general question: How good are bounds on pseudoscalar couplings to quarks? From ψ decays alone, one gets typically $x < 1/\sqrt{50}$. How much better can we expect from K decays?

The rate of K decays into axion-like particles has been strongly debated in the past. Let us distinguish between "mixing" and "hard" contributions. The first contribution arises from mixing of the light pseudoscalars with the pions, etc., ... While the contribution due to pion mixing can be estimated from the $K \to \pi\pi$ decay, we have no direct experimental insight for $K \to \pi\eta$, which is kinematically forbidden. Further uncertainty arises from the applicability of the $(\Delta I = 3/2)/(\Delta I = 1/2)$ suppression to these processes and from the possible interferences. Therefore, estimates for the branching ratio vary from 10^{-6} to 10^{-8} according to the various authors (for $x \sim y \sim 1$).

A "hard" contribution arises at the one-loop level.[4] The naive W-exchange graph initially proposed has to be supplemented by 12 other diagrams, which bring into play one more unknown, namely the mass m_H of the charged spin-zero boson(s) associated with our light pseudoscalar (see Figure 1).

The branching ratio is then computed to be:

$$B(K\to\pi h) = 0.8 \times 10^{-6} \{x A_1 (m_c) + x^3 A_2 (m_c) + (\frac{m_t}{m_c})^2 (S_2^2 + S_2 S_3) [x A_1(m_t) + x^3 A_2 (m_t)]\}^2 \tag{4}$$

where the numerical values of the functions A_1 and A_2 are plotted in Figure 2.

Barring accidental cancellations, we see that, for $m_H \geq 100$ GeV,

$$B(K\to\pi h) \sim 5\times10^{-4} x^2 \tag{5}$$

which would provide a bound $|x| < 10^{-2}$ at the present experimental accuracy.

To close this section, we would like to conclude that K decays provide us with a very sensitive tool to explore light pseudoscalar couplings. However this tool suffers from two limitations: one is due to uncertainties arising from the understanding of "mixing" contributions, the other is linked to the unknown mass of charged "Higgs" bosons.

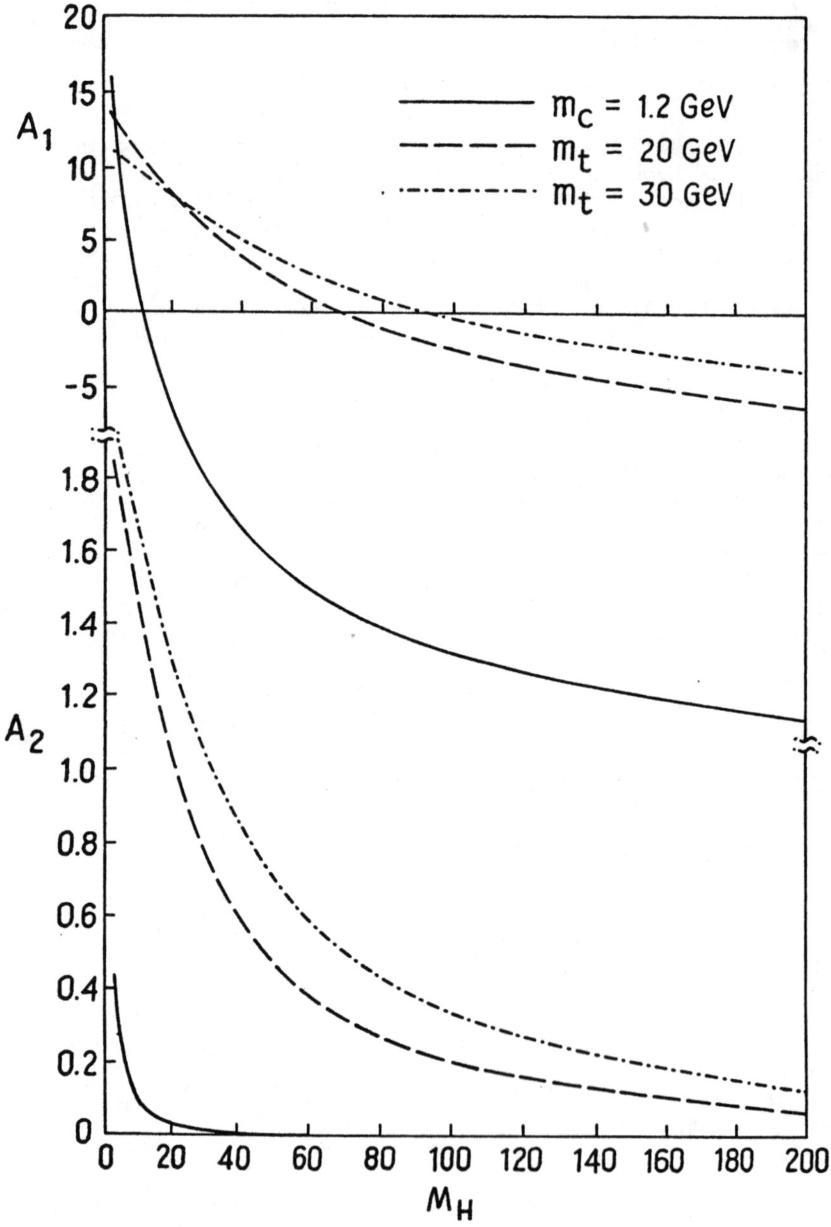

FIGURE 2

Numerical values of the functions A_1 and A_2 which determine $B(K\to\pi h)$ (see Eq. (4)).

The high sensitivity makes this tool quite unique, and places K decays in a prime position to establish the existence of such a coupling. Negative results unfortunately could most of the time be taken away be evoking (unlikely) cancellations between the various contributions.

FLAVOR NON-DIAGONAL COUPLINGS

We assume a "Goldstone" coupling of the kind (e.g. "familons",[5] technicolor, ...)

$$G\bar{s}\,\gamma_\mu\,\left(\frac{1-\gamma_5}{2}\right)\,d\,\partial^\mu h. \tag{6}$$

Limits on this coupling can be obtained somewhat indirectly and in a model dependent way (due to possible cancellations) from its contribution to the $K_L - K_S$ mass difference: one gets typically $G < 10^{-6}\ \mathrm{GeV}^{-1}$. Also, CP violation in K decays might be invoked; for an assumed phase of order one, we would get $|G| < 10^{-7.5}\ \mathrm{GeV}^{-1}$. Here we remark that $K \to \pi h$ fares much better since the limit derived from present experimental data gives:

$$G < 10^{-10}\ \mathrm{GeV}^{-1}.$$

(As an immediate consequence, such light bosons cannot be held responsible for CP violation!) Of course, the limits from $K\bar{K}$ mixing keep their value if h is too heavy to be produced in K decays.

REFERENCES

1. For a more detailed discussion, see for instance: J.-M. Frere, "Axions" Europhysics Study Conference on Unification of the Fundamental Interactions II, Erice 6-14 Oct. 1981, J. Ellis and S. Ferrara Editors (Plenum press) and references therein.
2. J. Ellis, D.V. Nanopoulos, P. Sikivie, Phys. Lett. 101B, 387 (1981).
3. J.L. Lee-Franzini in Proceedings, 21st International Conference on High Energy Physics, Paris, July 1982, p. c3-95.
4. J.-M. Frere, M.B. Gavela and J.A.M. Vermaseren, Phys. Lett. 103B, 129 (1981);
   L.J. Hall and M.B. Wise, Nucl. Phys. B187, 397 (1981).
5. F. Wilczek, Phys. Rev. Lett. 49, 1549 (1982) and these proceedings.

REPORT FROM THE HYPERON PHYSICS SUBGROUP

J.M. Frere
University of Michigan, Ann Arbor, MI 48109

D.A. Jensen (Chairman)
University of Massachusetts, Amherst, MA 01003

A.N. Kamal
University of Alberta, Edmonton Alberta T6G 2J1 Canada

J. Lach
Fermi National Accelerator Laboratory, Batavia, IL 60510

R.E. Mischke and T. Oka
Los Alamos National Laboratory, Los Alamos, NM 87545

ABSTRACT

A number of topics were considered in the Hyperon Group. As the time was reasonably short, and the group was not large, the list of topics is by no means exhaustive. The topics discussed should rather be considered a sampling of a much more extensive list of topics that might have been discussed. This report will thus consist of some brief remarks relevant to the topics discussed, and a few concluding remarks. The interested reader is referred to the references for appropriate details.

RIGHT-HANDED CURRENTS?

The standard picture of weak interactions is based on left-handed currents, i.e., a pure V-A structure. It is appropriate to ask to what extent may right-handed currents be ruled out. Beg[1] et al. in 1977 suggested that the V+A current terms could be as large as 13% of the V-A terms! More recently, Bigi and Frere[2] have suggested that one may set stringent limits on $\sin\xi$, the amount of right-handed charged current mixing in $\Delta I = 3/2$ non-leptonic hyperon decays, by studying this $\Delta I = 3/2$ component of weak hadronic hyperon decays alone. In these decays, there is a strong enhancement of a left-right interference term leading to much improved sensitivity to right-handed currents. Bigi and Frere determined $|\sin\xi| < 0.01$ as an upper bound.

The question of right-handed current effects in semileptonic decays has also been addressed.[3] It was pointed out that the long-standing discrepancies between the value of g_1/f_1 obtained from spin asymmetry measurements and Dalitz plot measurements in $\Lambda \to pe\nu$ can be resolved by the inclusion of significant amounts of right-handed current. In fact, the fits presented in Reference 3 suggest a right-handed component in the $\Delta S = 1$ current of almost 30%! It might be concluded that additional clarification of this fundamental question in theoretical and/or experimental regimes is sorely needed.

0094-243X/83/1020160-03 $3.00 Copyright 1983 American Institute of Physic

RARE HYPERON DECAYS, INCLUDING γ DECAYS

It must be clearly stated at the outset when discussing rare hyperon decays that data is woefully meager at best![4] First, consider the radiative decays. Excluding $\Sigma^0 \to \Lambda\gamma$, there are only two branching ratios: $\Sigma^+ \to p\gamma$ $(1.2 \pm .13) \times 10^{-3}$, and $\Xi^0 \to \Lambda\gamma$ $(5\pm0.5)\times10^{-3}$, and one asymmetry, $\alpha_{\Sigma p\gamma} = -.72 \pm .29$, that have been measured. There are also limits on $\Xi^0 \to \Sigma^0\gamma$ and $\Xi^- \to \Sigma^-\gamma$ (< 0.07, $< 1.2 \times 10^{-3}$ respectively).

There are three obvious diagrams that can contribute: a) single quark, b) two quark and c) three quark diagrams.[5] The data are inconsistent with the most simple assumption that baryon radiative decay proceeds exclusively via a single quark diagram. A somewhat more ambitious model including single and double quark diagrams can fit all of the (meager!) data. It is abundantly clear that in the area of radiative decays, data is certainly lacking.

One might also test invariance principles and conservation laws by looking for forbidden hyperon decays. For example, $\Delta S > 1$ transitions might be looked for by searching for $\Xi^0 \to p\pi^-$. Naively, one might expect the branching ratio to be $\sim 10^{-10}$. The current limit[4] is $< 3.6 \times 10^{-5}$. 10^{-10} seems very hard indeed! Other models such as $\Xi^- \to ne^-\nu$ and $n\mu^-\nu$ have also been sought, but the limits are not as stringent as for $\Xi^0 \to p\pi^-$. The study of Ω^- decays presents the opportunity to search for $\Delta S = 3$ decays! Of course, in the context of the standard models, a decay such as $\Omega^- \to n\pi^-$ would be most unlikely.

Searches for $\Sigma^+ \to p^+e^+e^-$, $\Sigma^+ \to p^+\mu^\pm e$, ... or $\Sigma^+ \to ne^+\nu$, ... to search for family, lepton number, or $\Delta S = \Delta Q$ violation must also be pursued to more constraining levels.

HYPERON BETA DECAY

Recent fits to the hyperon semileptonic decay rates and the value of g_1/f_1 may, if $\Sigma^- \to \Lambda e^-\nu$ is excluded, be well fit to an extended Cabibbo hypothesis with some SU(3) breaking[6], (extended in the Kobayashi-Maskawa sense.) It must be noted however, that except for n and Λ beta decays, the data input to the fits are only determined to $\sim$ 10%, or worse. Rates for only two muon modes ($\Lambda \to p\mu^-\nu$, $\Sigma^- \to n\mu^-\nu$) have been determined. It might be guessed that were there tighter constraints from the data, the business of doing "Cabibbo fits" would become very demanding. This guess is supported by recent efforts at doing detailed fits to $\Lambda \to pe\nu$ decays.[7] In these fits, it is found that the values of the form factors obtained depend, for example, on the form of q^2-dependence assumed.

Further, one finds that it is possible to calculate form factors within the context of the quark model or the bag model.[8] These form factors may be used as some of the input to fits when equivalent data are not available. Including the calculated value of the form factor g_2 (weak electric form factor), induces substantial changes in g_1 (axial vector form factor). It thus becomes clear that the high precision hyperon experiments are no longer simply weak interaction experiments, but are, in fact, experiments that probe the details of hadron structure.

CONCLUSION

The area of hyperon physics continues to be a rich field of study for theorists and experimentalists alike. As the experiments improve, the field of hyperon physics expands out of its traditional role as a testing ground for the Cabibbo theory of weak interactions and into the much fuller role of probing the details of hadron structure as well as providing a testing ground for the standard electroweak model. This expanded role is seen in each of the areas mentioned above. It is to be expected that the future of hyperon physics will continue to be both challenging and exciting.

REFERENCES

1. M.A.B. Beg, R.V. Budny, R. Mohapatra, A. Sirlin, Phys. Rev. Lett. 38, 1252 (1977).
2. I.I. Bigi, J.M. Frere, Phys. Lett. 110B, 255 (1982).
3. T. Oka, "Right Handed Current Effects in $\Delta S = 1$ Semileptonic Decays, Los Alamos LA-UR-83-618.
4. Particle Data Group, Phys. Lett. 111B (1982).
5. A.N. Kamal, R.C. Verna, Phys. Rev. D26, 190 (1982).
6. J.F. Donoghue and B. Holstein, Phys. Rev. D25, 2015 (1982).
7. J. Wise et al., Phys. Lett. 98B, 123 (1981);
   D. Jensen - private communication.
8. J.F. Donoghue and B. Holstein, Phys. Rev. D25, 206 (1982).

HYPERON DECAYS AND A LIMIT ON LEFT-RIGHT MIXING

J.M. Frere
Randall Laboratory of Physics
University of Michigan, Ann Arbor, MI 48109

I.I. Bigi
Rhein.-Westf. Tech. Hochschule
D-5100 Aachen, Federal Republic of Germany

Left-right symmetrical models, based on the group $SU(2)_L \times SU(2)_R \times U(1)$ have often been discussed as a somewhat minimal extension of the standard group $SU(2)_L \times U(1)$. As far as low-energy charged current gauge interactions are concerned, we need to introduce two new parameters, namely the mass of the hypothetical right-handed boson, M_R and its mixing with the left-handed boson, characterized by $\sin\xi$ which lead respectively to effective $(V+A)(V+A) \cdot \frac{G_F}{\sqrt{2}} \cdot M_L^2/M_R^2$ and $(V+A)(V-A) \cdot \frac{G_F}{\sqrt{2}} \cdot \sin\xi$ four-fermion interactions.

Existing limits on these two parameters usually rely on the measurement of leptonic polarization. However, in the popular case where right-handed neutrinos possess a large Majorana mass, making them impossible to produce at current accelerator energies, the leptonic charged current is effectively purely left-handed. Some information on M_R may be gained from the study of neutral currents, assuming a specific Higgs structure; this, however, gives no handle on the parameter ξ. One thus has to probe directly the charged hadronic current.

We have suggested that a Majorana mass-independent limit on ξ be obtained by considering deep inelastic $\bar{\nu}$ scattering at large values of x and y (various Q^2 should be considered to eliminate potential higher twist contributions). The limit obtained in this way by the CDHS group is independent of any Majorana mass and gives[1] $|\sin\xi| <$ 10%. On the other hand, more speculative, but more sensitive tests can be extracted from the consideration of non-leptonic decays. Obviously the picture is obscured here by the presence of strong interaction corrections to the g_A coupling (Adler-Weisberger relation).

A limit can, however, be obtained by isolating some special channels and kinematical regions where contribution from LR mixing would be considerably enhanced. This is found to be the case for the $\Delta I = 3/2$ amplitude of hyperon decays. We have shown that, taking into account chiral considerations and radiative corrections enhances the left-right contribution in P-conserving decays by a factor of 120, leading to

$$\frac{A^L(\Delta I=3/2) + A^{LR}(\Delta I=3/2)}{A^L(\Delta I=3/2)} \sim 1 - 120 \sin\xi$$

0094-243X/83/1020163-02 $3.00

for p-wave contributions, while s-wave contributions are unaffected if L and R Cabibbo mixings are taken to be equal.

Comparison with present data on these amplitudes then suggests $|\sin\xi| < 1\%$ as a reasonable limit.[2]

REFERENCES

1. H. Abramowicz et al., Zeit, fur Phys. C12, 225 (1982).
2. For more details, see I.I. Bigi and J.-M. Frere, Phys. Lett. 110B, 255 (1982) and references therein.

NEUTRINO PHYSICS WITH INTENSE BEAMS

Boris Kayser
National Science Foundation, Washington, D. C. 20550

S. Peter Rosen
National Science Foundation, Washington, D. C. 20550
and
Purdue University, Lafayette, Ind. 47907

ABSTRACT

Intense, low-energy neutrino beams would make possible a variety of interesting experimental tests of the standard model. A number of these tests are described.

The neutrino subgroup (D. Bryman, R. Carlini, M. Lusignoli, G. Stephenson, P. Vogel, and the authors) concludes that neutrino experiments with intense low-energy neutrino beams can provide interesting tests of the predictions and underlying assumptions of the standard model. The experiments will complement those expected to be carried out on the Z^0 resonance, yielding information which the latter experiments cannot provide.

A number of the potential low-energy experiments involve neutrino-electron scattering. To achieve precision measurements in this reaction, it is necessary to use an intense narrow-band beam (NBB). Because of kinematics, the "narrow" requirement limits one to small angles and long, "skinny" detectors, and rate is lost by virtue of the inverse square law. NBB detectors are therefore likely to be small, of the order of 100 tons. Annual event rates in such detectors as a function of primary proton current are shown in Table I.[1]

Table I. Neutrino event rates with a 100-ton detector, and a 10%-width narrow-band beam at 1.5 GeV

30 GeV Primary Proton Current	Total # $\nu_\mu e \rightarrow \nu_\mu e$ 12 Months	
[μa]	BNL (Horn)	Ideal Focussing Horn
1	7	36
10	70	360
100	700	3,600

0094-243X/83/1020165-05 $3.00 American Institute of Physics

Potential low-energy high-intensity νe experiments which are felt to be particularly interesting are these:

1. MEASUREMENT OF $SIN^2\theta_W$ TO $\lesssim$ 1% ACCURACY IN $\overset{(-)}{\nu}_\mu e \rightarrow \overset{(-)}{\nu}_\mu e$

Given a value of $\sin^2\theta_W$ measured in low-energy ($E \ll M_W$) reactions, one can apply ~5% radiative corrections predicted by the standard model[2] to it, and predict M_W. If the latter is determined directly with an accuracy of ~1% before the intense low-energy neutrino source is completed, the new low-energy $\sin^2\theta_W$ measurements will permit a test of the radiative corrections. The accuracy of the $\sin^2\theta_W$ value (more precisely, the $\sin\theta_W$ value) should match that of M_W, which requires going beyond the accuracies currently being achieved. (The theoretical uncertainty in the M_W prediction for a given $\sin^2\theta_W$ is small compared to 1%.) The $\overset{(-)}{\nu}_\mu e$ reaction is advantageous because it is free of hadronic complications.

Before long, the Z^0 will probably be discovered and M_Z determined more accurately than M_W. From $\sin^2\theta_W$ measured at low energies, radiative corrections, and standard model assumptions about ρ (the ratio of neutral-current to charged-current interaction strengths), one can predict M_Z. Thus, a test of this prediction is a probe of possible surprises in the ρ parameter.[3]

2. CROSS-SECTION MEASUREMENTS

Measurements of the total cross-sections σ_{Tot} ($\overset{(-)}{\nu}_\mu e \rightarrow \overset{(-)}{\nu}_\mu e$) and the differential cross-section

$$\frac{d\sigma}{dy} = A + 2B(1-y) + C(1-y)^2$$

to 1% will provide several pieces of fundamental information. The σ_{Tot} measurements will determine the parameter ρ, and a comparison of its value with the measured values of M_W, M_Z, and $\sin^2\theta_W$ through the formula

$$\rho = \frac{M_W}{M_Z \cos\theta_W}$$

will check the usual assumption that all the Higgs scalars in the standard model belong to weak isodoublets.

A limit on the B-coefficient in the differential cross-section will set a limit on the presence of anomalous Lorentz structure in the neutral-current interaction. It is also important to note that detection of a positive value for B will establish that ν_μ is a Dirac neutrino and not a Majorana one;[4] values of $B \leq 0$ do not distinguish between the two types of neutrino.

3. STUDY OF THE REACTION $\overset{(-)}{\nu}_e e \rightarrow \overset{(-)}{\nu}_e e$

At an accelerator which serves as a K-factory, one can produce neutrino beams enriched in <u>electronic</u> neutrinos by using the neutrinos from $K_L \rightarrow \pi e \nu_e$ or from a muon racetrack. The $\overset{(-)}{\nu}_e$ enriched beams make possible a number of interesting experiments, including study of the the interference between neutral-current (NC) and charged-current (CC) contributions in $\overset{(-)}{\nu}_e e$ scattering.[5] The $\overset{(-)}{\nu}_e e$ reaction is one of the few places where the sign of a neutral-current amplitude relative to that of <u>another kind</u> of amplitude may be probed. To confirm that in $\overset{(-)}{\nu}_e e$ scattering the NC term, assumed present, is negative relative to the dominant CC term, as predicted theoretically, requires an accuracy of $\sim$25% in the cross section measurement.[6] To <u>demonstrate</u> that the NC term is indeed present requires an accuracy of $\sim$10%. Such a demonstration would show that the NC interaction of neutrinos is (at least partially) flavor preserving.[7]

We turn now to interesting possible neutrino-hadron scattering experiments.

4. TEST OF $\nu_e - \nu_\mu$ UNIVERSALITY

The existence of a ν_e-enriched beam will enable one to check the universality of neutral-current neutrino couplings. In particular one can compare the inclusive cross-sections for $\nu_e + N \rightarrow \nu_e + X$ and $\nu_\mu + N \rightarrow \nu_\mu + X$ to compare the $\nu_e - Z^0$ and $\nu_\mu - Z^0$ couplings at the 10% level.

5. DIRECT TEST OF SPIN STRUCTURE OF NEUTRAL-CURRENT INTERACTIONS IN $\nu N \rightarrow \nu N^*$

The V, A (fermion helicity preserving) character of both charged- and neutral-current weak interactions is a basic underlying assumption of the standard model. In spite of the voluminous data on neutral-current interactions of neutrinos, there is essentially no direct evidence at all that these interactions obey this assumption.[8] Such evidence could be obtained through a study of the forward- and backward-scattered neutrinos from the reaction $\nu N \rightarrow \nu N^*$, where N^* is the (3,3) resonance.[9] Conservation of angular momentum along the beam line, J_{beam}, implies that for forward-scattered neutrinos, $|J_{beam}(N^*)| = 1/2$ unless the neutrino helicity flips. For backward-scattered neutrinos, $|J_{beam}(N^*)| = 1/2$ unless the helicity <u>does not</u> flip. In the decay $N^* \rightarrow N\pi$, the pion angular distribution is very different for $|J_{beam}(N^*)| = 3/2$ than for $|J_{beam}(N^*)| = 1/2$, so it should be possible to draw conclusions about the behavior of the neutrino spin.

6. EXCLUSIVE SCATTERING PROCESSES

Once the precise structure of the neutral-current interaction has been determined, one can use exclusive neutrino-hadron scattering processes to probe both nucleon and nuclear structure[10] at small q^2, down to $q^2 \simeq 0.2\ \mathrm{GeV}^2$. Isoscalar and isovector, proper vector and axial vector form factors can be studied independently by picking out specific nuclear transitions. Comparison with proper vector form factors measured in electromagnetic transitions will provide important insights into collective effects and other aspects of nuclear structure.

7. NEUTRINO OSCILLATIONS

By operating the machine so as to give a high flux of low energy pions, one can obtain an intense source of low energy neutrinos. Such a source is ideal for neutrino flavor change or flux disappearance experiments that can push the limits on oscillation parameters down to $\sin^2 2\theta \lesssim 0.001$, $\Delta m^2 \lesssim 10^{-3}\ \mathrm{eV}^2$, and beyond.[11]

REFERENCES

1. Contributed by R. Carlini.
2. A. Sirlin and W. Marciano, Nucl. Phys. B189, 442 (1981), and C. Llewellyn Smith and J. Wheater, Phys. Lett. 105B, 486 (1981).
3. We thank W. Marciano for discussions of this point and the question of radiative corrections.
4. S. P. Rosen, Phys. Rev. Lett. 48, 842 (1982).
5. B. Kayser, E. Fischbach, S. P. Rosen, and H. Spivack, Phys. Rev. D 20, 87 (1979).
6. There is already some evidence that this term is negative from the reactor experiment reported in F. Reines, H. Gurr, and H. Sobel, Phys. Rev. Lett. 37, 315 (1976). However, confirmation is desirable.

7. For additional discussion of $\overset{(-)}{\nu}e$ scattering as a test of the standard model, see K. Abe, F. E. Taylor, and D. H. White, in *Proceedings of the 1982 DPF Summer Study on Elementary Particle Physics and Future Facilities*, Snowmass, Colorado, edited by R. Donaldson, R. Gustafson, and F. Paige, p. 165.

8. B. Kayser, G. Garvey, E. Fischbach, and S. P. Rosen, Phys. Lett. 52B, 385 (1974), and R. Kingsley, F. Wilczek, and A. Zee, Phys. Rev. D 10, 2216 (1974). For beautiful recent evidence that the charged-current weak interactions do preserve fermion helicity, see M. Jonker *et al.*, Phys. Lett. 86B, 229 (1979).

9. B. Kayser, "Neutral Weak Interactions and the Possibility of Right-Handed Neutrinos," based on a talk presented at the Annual Meeting of the American Physical Society, 1975 (unpublished).

10. T. W. Donnelly and R. Peccei, Phys. Reports 50, 1 (1979).

11. For a discussion of some present and future neutrino oscillation experiments, see H. Chen in *Proceedings of the Third Workshop on Grand Unification*, Chapel Hill, 1982, edited by P. Frampton, S. Glashow, and H. van Dam, p. 206.

ON THE FEASIBILITY OF PERFORMING NEUTRINO EXPERIMENTS AT AN INTENSE SOURCE OF STRANGENESS

R. D. Carlini and G. J. Stephenson, Jr.
Los Alamos National Laboratory

Abstract

Several very important low energy neutrino experiments have been identified and discussed by Kayser and Rosen in these proceedings. We address the possibility of performing some of these experiments in the environment of a kaon factory. To carry out specific analyses, we draw upon the design of existing BNL detectors, the design of experiments that are proposed for an 800 MeV proton driven facility at Los Alamos, and the use of the code NUBEAM developed at CERN. Since in most cases we have not optimized, this is a conservative estimate. We find that many of the desired experiments can be performed with facilities now under discussion.

Introduction

Several important experiments that could be performed with beams of low energy neutrinos were considered by the neutrino subgroup. Boris Kayser and Peter Rosen[1] have prepared an excellent summary of the physics issues, identifying seven major areas of interest. In view of those interests, the conference organizers have asked us to provide a discussion of the feasibility of such experiments at an intense source of low energy neutrinos that one would expect to have in conjunction with a kaon factory. It is not possible for us to respond in the detail that we would wish, nor to discuss the entire range of experiments identified in the article of Kayser and Rosen. However, we hope that we are providing enough detail to convince the reader that a range of important neutrino experiments is feasible. These experiments will provide unique information about the structure of fundamental interactions and therefore constitute an additional justification for the construction of such a facility.

In order to truly discuss the feasibility of any neutrino experiment it is necessary to specify both the neutrino source and the neutrino detector. The specification of the source includes the energy and intensity of the primary proton beam, the construction of the proton target, and the focussing device (presumably a horn) to focus the mesons for increased neutrino flux and, where possible, for neutrino type selection. These factors translate into a neutrino spectrum over the detector. A detector must be specified in order to translate the neutrino flux into a counting rate, and to estimate the kinds and amounts of backgrounds, and to assess the ability of a particular detector to suppress those backgrounds.

In this article, we shall rely upon comparisons with the existing program at Brookhaven and with detailed designs included in the Los Alamos proposal for an intense neutrino source using the existing LAMPF beam of 800 MeV protons as the primary beam on the production target[2]. We also employ the Monte Carlo code NUBEAM that was developed at CERN[3]. This code reproduces the operating conditions at the Brookhaven AGS and the results from the independent Monte Carlo code LANLNU[4] that is used in the Los Alamos neutrino facility proposal[2]. Also, a code based on the SLAC EGS Electromagnetic shower code[5] was used to model the detector.

Turning first to the question of neutrino fluxes and relative counting rates, we have chosen to use $\nu_\mu e \rightarrow \nu_\mu e$ scattering to compare rates, a nominal 100 micro-amps on target and various horn designs. To specify the calculation of the neutrino fluxes, we refer to Fig. 1. For the low energy case, a horn shape in the decay tunnel is assumed as shown on Fig. 2. For the distribution of pions produced by 800 MeV protons, we believe that the shape is nearly optimal, although some tens of per cent increase may yet be achieved. The variation of neutino flux with decay channel length is shown on Fig. 3 and the neutrino spectrum is shown on Fig. 4.

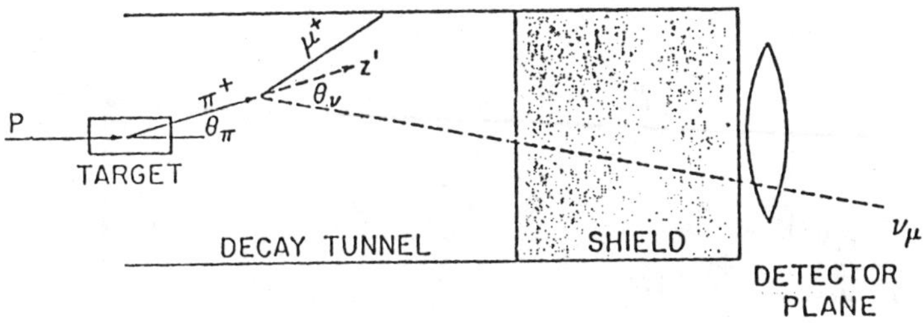

Fig. 1. General facility geometry.

For the case of 30 GeV protons, we have used the present BNL two horn configuration. For this configuration, the variation in total neutrino flux and in mean neutrino energy with the length of the decay tunnel are shown on Fig. 5 and the neutrino spectrum is shown on Fig. 6.

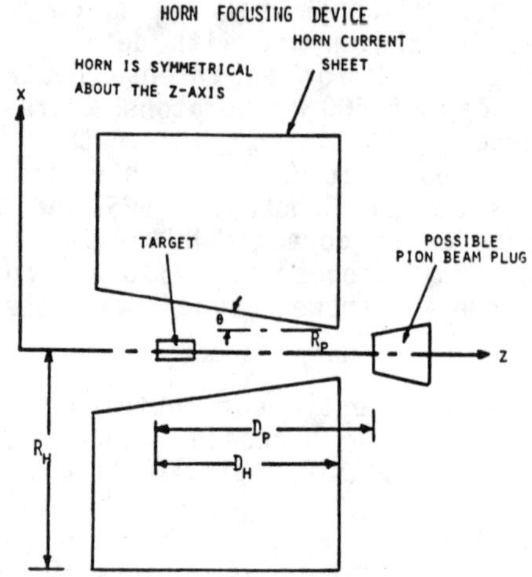

Fig. 2. Focussing horn for 800 MeV protons.

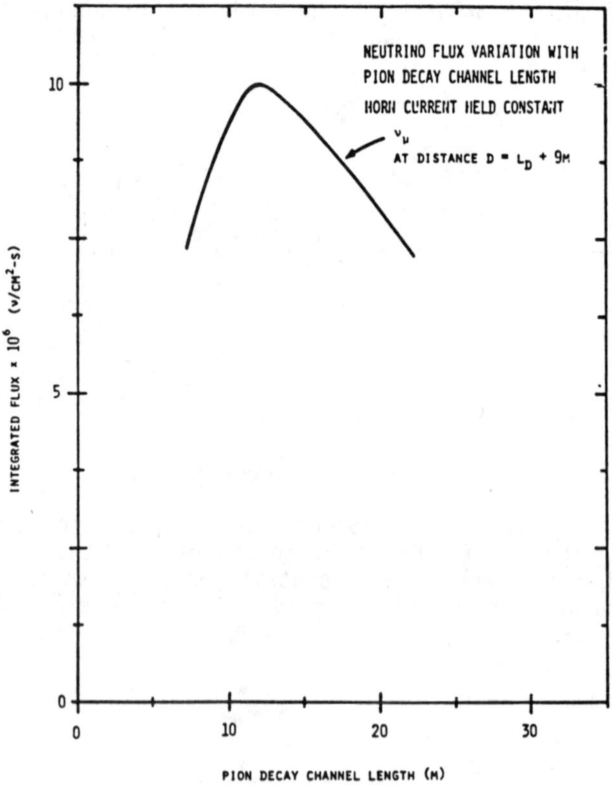

Fig. 3. Flux variation with decay channel length, 800 MeV proton.

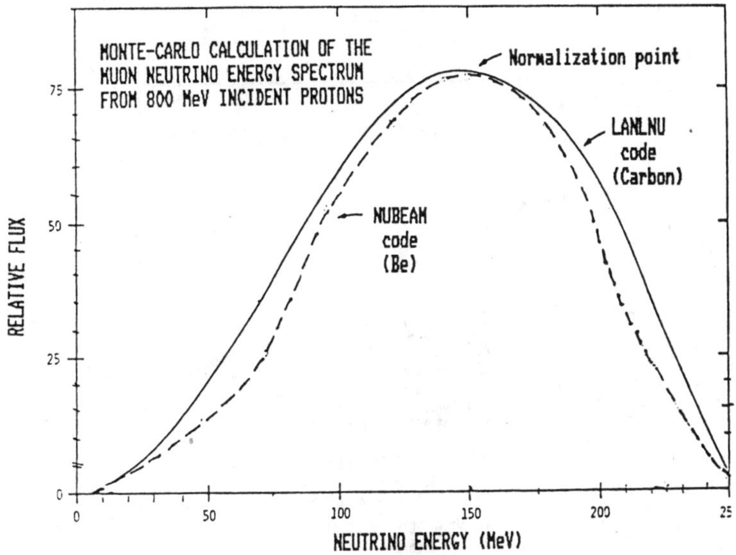

Fig. 4. Neutrino energy spectrum for 800 MeV protons.

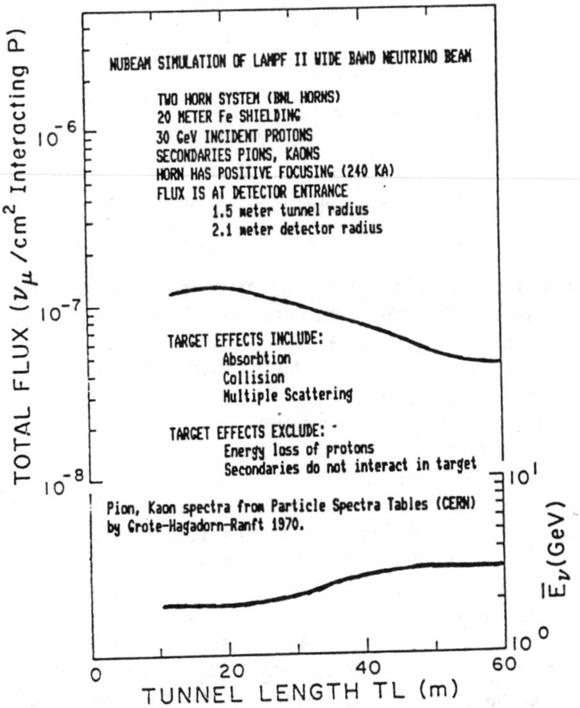

Fig. 5 Variation of total neutrino flux and the mean neutrino energy with decay tunnel length for 30 GeV protons.

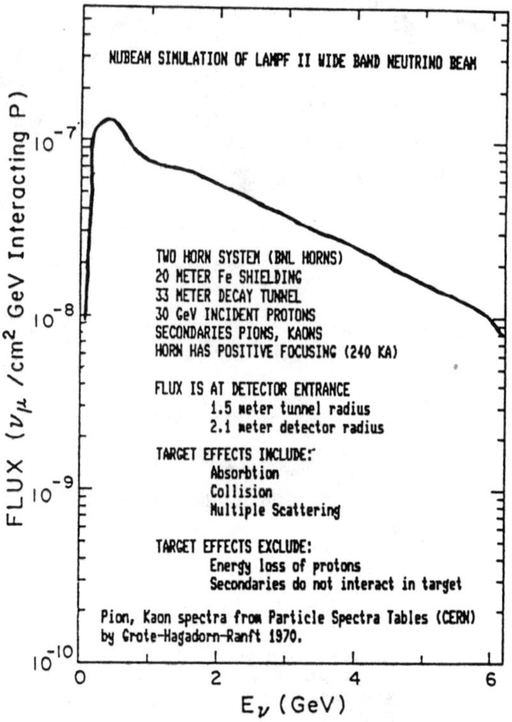

Fig. 6. Neutrino energy spectrum for a wide band beam and 30 GeV protons.

Turning now to counting rate estimates, for the 800 MeV proton case we use the detector design of the Los Alamos National Laboratory proposal.[2] To be more specific, the detector is shown schematically on Fig. 7. Its parameters are:

Size = 5-m x 5-m x 22-m
Total aluminum tonnage = 366 tons
Total scintillator tonnage = 136 tons
Average density = 0.91 gm/cm^3
Average Rad. length = 38 cm (over all detector)
Average dE/dx = 1.6 MeV/cm (minimum ionizing)
Average gamma - conversion length = 49 cm
Per cent electron energy deposited in scintillators = 30% (visible total energy)

The pions are horn focussed as discussed in the Los Alamos proposal. Although the focussing is wide band, the energy spectrum of interacting neutrinos tends to mimic a narrow band beam. This <u>interacting</u> spectrum is roughly ± 30% wide.

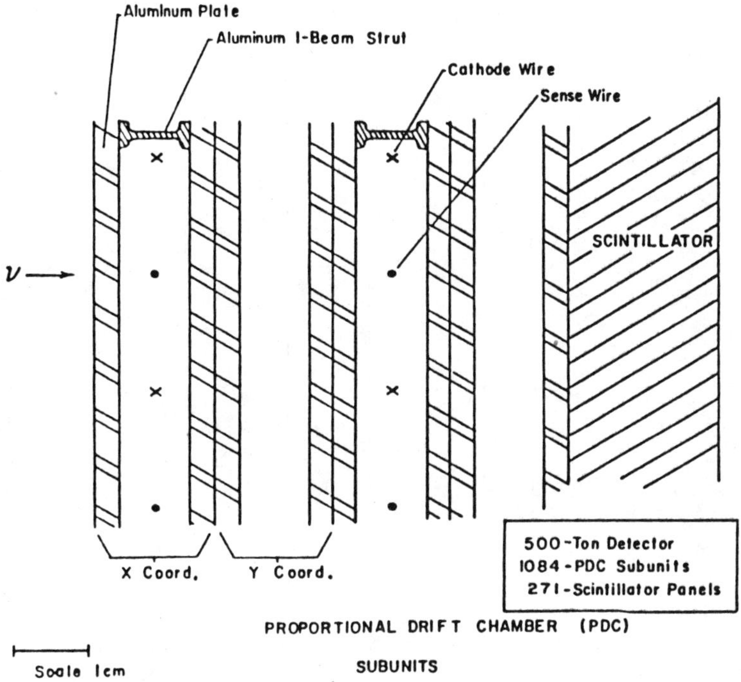

Fig. 7 Schematic of a module of the detector used for rate estimates with 800 MeV protons.

To make estimates for the case of 30 GeV protons, since the counting rates increase with the neutrino energy and beam associated backgrounds originating in inert material also increase dramatically, we eliminate the inert material, resulting in a design that essentially matches the existing Penn-Brown-Tokyo-BNL detector. For narrow band beam counting rate estimates, we have assumed narrow band beam focussing of the pions with a mean energy of 1.5 GeV and a spread of $\pm$ 15%. This is a very conservative narrow band beam design.

In the high energy case, an electron energy cut is set at 200 MeV; for the low energy case the cut is at 75 MeV. The data are summarized in Table I.

As one can see from the rates quoted in the last column, there is the possibility of a great gain in statistics over the LAMPF facility in going to a kaon factory. However, one must consider also background supression and the ability to extract the relevant

		$<T_p>$ GeV	ν_μ Flux $\#/cm^2$-s	Narrow Band Horn Efficiency	$<E\nu_\mu>$ GeV	Relative $\nu_\mu e \rightarrow \nu_\mu e$ Rate	Conditions
BNL Wide Band Horn	scaled from measurement	30	3×10^7	----	1.98	----	BNL Geometry
	NUBEAM		1.58×10^7				
Estimated BNL Narrow Band with NUBEAM		30	1.1×10^6	1/15[a]	3	4.8	BNL Geometry
LAMPF II Ideal[b] Wide Band (0°-45°)		30	2.4×10^8	----	.94	----	50 m decay tunnel 20 m Fe shield
LAMPF II Ideal[b] Narrow Band (0.57°-3.44°)		30	1.5×10^7	1/15	1.56	32	"
LAMPF II experimental area: NUBEAM simulation		30	4.8×10^7	----	2.5	80	"
Los Alamos Neutrino Facility Proposal	LANLNU	0.8	1.0×10^7				12 m decay channel
	NUBEAM[c]		1.87×10^7	----	.16	1	9 m Fe shield

a Based on experience and estimates for improved BNL two horn system.

b Reference 6 with target effects included.

c Low energy target effects (proton and pion energy loss) not included. This is 2x two high. LANLNU results taken as normalization.

Table I - Fluxes and relative ν_μ-e elastic count rates.

physics. It is worth noting that the $\langle E_\nu \rangle$ from a kaon factory can be tuned to a lower energy if a specific experiment requires it. However, beam associated backgrounds are a function of the proton energy as well as the neutrino beam characteristics. These backgrounds include high energy neutrons, which necessitate additional shielding between the source and the detector, and electron neutrinos produced in time by kaon decays. Many methods of cleaning up the neutrino beam will involve some decrease in the useful flux, so a neutrino physics program will benefit from the highest possible proton currents. For some specific experiments variable proton energy will be useful.

As Kayser and Rosen point out, $\nu_\mu e$ scattering is most interesting if one can measure a y-distribution. To do so one requires a measurement of the energy of the scattered electron and enough statistics to regress out the shape of the interacting neutrino spectrum.

In Fig. 8 we show a Monte Carlo reconstruction of y-distributions for the 800 MeV proton $\langle E_\nu \rangle$ = 170 MeV case and for the 30 GeV proton $\langle E_\nu \rangle$ = 1.5 GeV case. In the latter example the narrow band beam focussing condition described above is assumed. Based on current BNL data, π° backgrounds appear tractible and the random error per event is greatly reduced in the narrow band case. Some complimentary angle information is also available if the detector has sufficiently fine resolution.

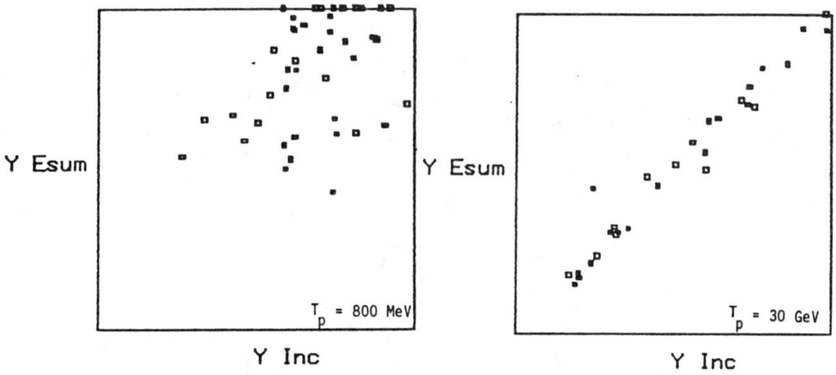

Fig. 8. Y reconstructed by energy only vs. Y actual.

Quoting from the Los Alamos National Laboratory proposal, with 100 micro-amps of proton current and 500 tons of detector, one should obtain a 1 per cent ($\pm$ 0.002) measurement of $\sin^2 \theta_W$ in 100 running days.

Another measurement that was singled out for study is the exclusive ν_μ to ν_e oscillation channel at low mixing probability as well as small Δm^2. At 800 MeV, that will be limited by statistics to a few times 10^{-6}. To achieve such a low level, it is necessary not only to suppress cosmic ray backgrounds, but to take advantage of the lack of beam related π°'s at those neutrino energies. It will also be necessary to subtract the electron events from electron neutrinos produced in π_{e2} decay. The latter will be possible since the spectrum of ν_μ peaks at over twice the energy of the ν_e spectrum.

For small amplitudes, the signal can be dominated by ν_e's from π_{e2} decay (BR ~ 10^{-4}). Then counting rate becomes crucial, if backgrounds are under control, since limiting values improve as the fourth root of the total counts. For this study, an optimum neutrino energy may well be below the maximum achievable at a kaon facility and a narrow band beam will be crucial.

Several neutrino-nucleus experiments were also singled out for study. On Fig. 9 we display inclusive cross-sections for charged current reactions on Carbon as a function of neutrino energy[7]. Clearly counting rates can be expected to rise as the mean

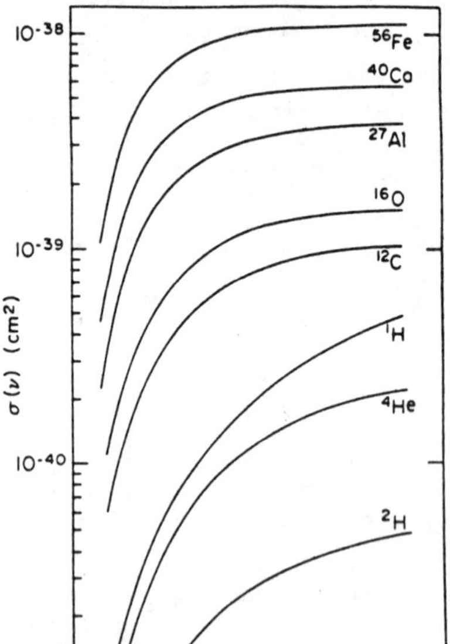

Fig. 9. Total neutrino cross sections σ(ν) versus neturino energy ν using the standard model.

neutrino energy rises above 200 MeV. Backgrounds related to a particular exclusive signal (a high energy decay gamma ray, for example) must be considered in establishing the optimum energy of the neutrino beam. In fact, different energy dependencies of signal and background may be necessary to measure exclusive cross-sections with sufficient accuracy to use the nucleus as a filter for spin and isospin characteristics of the weak currents.

Since background suppression always depends on the detector and since targets and detectors are one in neutrino physics, we cannot discuss real sensitivities here. In Table II, we simply list the counting rates estimated for 800 MeV protons with the detector described above (Fig. 1) and the neutrino flux of Fig. 5. These counting rates will change with neutrino energy, flux, detector composition and efficiency and are given here as illustration of the possibility of performing such experiments.

$N(\mu^- N^{12})$	195 events/ day
$N(\mu^+ B^{12})$	7 events/day
$N(\nu_\mu C^*)$	114 events/day

Table II - Selected counting rates.

It is clear to us from the discussion above that the physics described by Kayser and Rosen[1] is definitely accessible with a low energy neutrino facility that can be part of any medium energy intense source of kaons and should form a substantial part of the justification for such a facility.

References

1. Boris Kayser and S. Peter Rosen, these proceedings.
2. A Proposal to the Department of Energy for a National Facility to Provide a High Intensity Neutrino Source, Los Alamo National Laboratory, December 31, 1982. For copies or information contact the Working Group Chairman, G. J. Stephenson, Jr., Physics Division, MS D443, Los Alamos National Laboratory, Los Alamos, NM 87545.
3. C. Visser, "NUBEAM," HYDRA Application Library, CERN (1979).
4. T. Dombeck, "LANLNU--A Description of the Monte Carlo Calculation of the Neutrino Flux at the High-Intensity Los Alamos Neutrino Source," Los Alamos National Laboratory Report LA-UR-82-1589, May 1982.
5. R. L. Ford and W. R. Nelson, "The EGS Code System: Computer Programs for the Monte Carlo Simulation of Electro-Magnetic Cascade Showers, Version 3," SLAC-210-UC-32 (June, 1978).

6. J. C. Wang, R. C. Allen, and H. H. Chen, "Possible Neutrino Beams, Fluxes, and Rates at LAMPF II," Physics with LAMPF II, T. S. Bhatia, R. D. Carlini, R. R. Silbar, H. A. Thiessen, eds., Los Alamos Report LA-9798-P (1983).

7. T. W. Donnelly, "Intermediate-Energy Elastic Neutrino Scattering form Nuclei," Proceedings of the Los Alamos Neutrino Workshop, F. Boehm and G. J. Stephenson, Jr., eds. Los Alamos Report LA-9358-C (1982).

WHAT LIMITS EXPERIMENTS IN RARE KAON DECAYS: MACHINES/BEAM LINES OR DETECTORS?*

D. E. Dorfan and H. F.-W. Sadrozinski
University of California, Santa Cruz, California 95064

and

W. C. Louis III (Chairman)
Princeton University, Princeton, New Jersey 08544

There is a good possibility that, in ten years, we may have available Kaon beams with intensities a factor of 100 higher than present beams. This invites experimenters to confront the question of whether we will be able to make use of this increased intensity to reach lower branching ratios, or whether we are presently or will be limited by other factors. The participants in the "Detectors" subgroup posed the question: What limits experiments in rare Kaon decays -- machines, beam lines or detectors?

Under certain circumstances the question posed could have a simple answer. For example, if the present state of technology were such that in a ten year run no decays of the desired mode would be expected, even if a beam could be designed which transported every Kaon produced to a detector with a 100% acceptance for decays, then the answer would be: machine. At the other extreme one could envisage a situation where the machine and beam line were such that a pencil beam of pure Kaons with an arbitrary small momentum bite could be produced with a flux so large that the detector was unable to handle the singles rates due to K decays in the decay volume alone. Here the answer would obviously be: detector.

These examples are clearly ludicrous, but serve to highlight the fact that the situation is complex, and in practice tradeoffs between beam intensity, cleanliness, momentum bite and beam size are needed. Thus one can make a case for any of the factors depending on how one views these tradeoffs. In fact, it seems more appropriate to ask a different question, namely: "How difficult (costly) is it to set a limit on rare processes, given a primary proton beam of specified energy, intensity and duty factor?" Obviously no answer to the latter question could be produced in the time available. Furthermore, we should remember that it is necessary to guess at technical advances in detector capabilities which will occur in the time before a new machine could deliver beam.

This report, therefore, has more modest aims: to define the factors which bear on the question; to describe the current status of these factors; and to illustrate them by listing the presently perceived limitations for specific decay modes. The reader will also find detailed background information on existing machines, beam lines and specific experiments in the papers of L.S. Littenberg and B. Winstein in these proceedings.

*This is a compendium based on several individual contributions, as well as the efforts of the chairman.

FLUX

Given a flux of N Kaons/sec, a running time of t seconds, a detection efficiency η and a decay volume such that a fraction f of the incident Kaons decay, the branching ratio, b, which would be expected to produce a single decay is given by:

$$b = (Nf\eta t)^{-1}.$$

Ideally $f = \eta = 1$; with $t \approx 10^7$ seconds, $b = 10^{-7} N^{-1}$. Thus a flux of 10^6 Kaons/sec would yield an average of one event for $b = 10^{-13}$. In practice f and η are $\sim 10^{-1}$ for experiments involving K decays in flight. Thus a flux of 10^6 Kaons/sec would give one event for $b = 10^{-11}$. We will call this branching ratio the sensitivity of the experiment. What fluxes are currently available? At the AGS one could, in principle, get a K^0 beam with a flux of 10^9 K^0's/sec if one used a proton beam of 5×10^{12} protons/sec on target and a secondary beam with an acceptance of 100 μ steradians. This beam, if usable, would yield a sensitivity of 10^{-14}.

For K^+ beams at 6 GeV/c, a flux of $\sim 4 \times 10^7$/sec is attainable using 5×10^{12} protons/sec on target, yielding a sensitivity $\sim 4 \times 10^{-13}$.

For stopping K^+ beams, experiments are being proposed where $\sim 10^5$ K^+/sec will be stopped. Since this type of experiment has $f = 1$ and $\eta \sim 1$, the maximum achievable sensitivities would be limited to $\sim 10^{-12}$.

DETECTORS

We note that these sensitivities determined by the available flux are far better than any sensitivities achieved to date. There is of course a time lag since published results come from experiments using technologies which were available when these experiments were proposed. Nonetheless it is necessary to examine whether currently available detectors are capable of reaching these limits. The key factors discussed in this regard are: singles rates, trigger rates and physics backgrounds.

Singles Rates: A fundamental limitation on the rate to which drift chambers can be exposed is set by space charge effects which cause local losses in efficiency. The threshold for this inefficiency occurs at a flux $\sim 10^4$/sec/mm of wire. Assuming 1 cm wire spacing, this corresponds to $\sim 10^9$/sec/m^2 for a uniformly distributed flux. Of course one can in principle do better by using a narrower wire spacing and larger detectors. This is another illustration of the futility of trying to answer in generality the question posed by the title of this report.

In practice, detectors are run at lower fluxes than the limit set above since problems with track reconstruction in the presence of large numbers of unrelated tracks have been the dominant factor to date. Experience has shown that current detectors can handle singles rates ~$10^7 - 10^8$/sec (assuming 100% duty cycle), and progress in this area is expected in the next decade.

For an ideal K beam where the dominant contribution to the singles rate comes from K's decaying in the volume, this would limit the usable flux to $10^8 - 10^9$/sec, assuming 10% decay in the decay volume. In practice, because of beam impurities (neutrons, pions, protons), shielding problems etc., actual rates would exceed this ideal by a factor of 10. Thus singles rates limit usable Kaon fluxes to ~ 4×10^6 - 4×10^7 (40% duty cycle at the AGS).

Nonetheless one cannot conclude on the basis of the above numbers that the detectors are the limiting factor, since higher flux could be used to produce cleaner beams. However, there is a limit to this argument as the singles rate in the ideal beam is close to being a limiting factor.*

Trigger Rates: Triggering problems may well limit the sensitivity for a particular decay mode. This limitation can take two forms: one is the case in which a prevalent decay mode is so similar to the desired rare mode that a trigger which separates them cannot be devised; secondly, randoms can play a part in triggering, thus producing an unacceptably high trigger rate.

The first type is independent of beam flux and sets a limit on the integrated flux which can be used. This is one clear case where the blame can be laid squarely on the detectors.

False triggers caused by randoms are a more general problem and are related to instantaneous fluxes. Progress in the design of sophisticated triggers has been extremely rapid during the past few years and this is not expected to be as serious a limitation as the maximum tolerable singles rate. Further, since the tradeoffs in intensity vs. cleanliness are similar, the two problems can be studied together.

As smaller and smaller branching ratios are measured, false triggers tend to become the limiting factor. However, their level is virtually impossible to predict *a priori*, since more and more unlikely conspiracies become significant and the relative rates increase at least linearly with beam intensity. Experience has shown that achievable sensitivities tend to increase by about a factor of ten for each successive experiment aimed at measuring a particular rare process. A large part of this can be attributed to the experience gained in dealing with this type of background. In practice, the best way to deal with these systematics is to collect a large enough sample of events (preferably under varied running conditions, e.g., at various beam intensities) so as to be able to study distributions for trends. This of course means that measurable branching

---

*It is interesting to note in this context that K^0 decay experiments at SLAC in the early 1970's, using conventional wire chambers in a beam with a duty cycle of 2×10^{-4} were able to handle beams with instantaneous intensities of ~ 10^7 K^0's/sec. These beams were, however, essentially free of neutrons.

ratios will not be as small as the "sensitivities" which are quoted by at least a factor of ten. Conversely, it may be necessary to have an order of magnitude larger intensity available to achieve a given level of sensitivity.

CONCLUSIONS

The subgroup tried to address the question of what limits rare decay experiments. They found that even a modified form of the question could not be answered in general because of the (mode specific) problems involved in estimating backgrounds. In addition, it is very difficult to make conclusive predictions in the face of uncertain projections regarding detector technology. However, studying the problem is still useful as certain hard limits (at least with current detectors) may be emerging.

One potentially interesting approach which may provide intuitive guidance would be to study the general subject of rare decays historically, examining in detail how available beam fluxes and sensitivities have evolved in the past. An example of this can be found in the contribution of R.E. Mischke in these proceedings on limitations for $K \to \pi\nu\bar{\nu}$ experiments. The study need not be limited to Kaon decays. Pion beams of high intensity are currently available. It would be an interesting and instructive exercise to step back in time and try to answer the questions posed here while limiting oneself to the facts known at that time.

As evidenced in the papers in these proceedings by Littenberg, Mischke and Winstein, the subgroup participants did agree that, on one hand, current machines have not been used to their potential limits. On the other hand, in many cases it could be enormously easier to reach a given sensitivity level by using a higher intensity machine. Whether or not one can go further could not be definitively answered in the absence of the time and means to perform large exercises such as those suggested here.

Finally, we note that in the past, new ideas which could not have been foreseen have invariably emerged as a result of further experience. Future experiments to study rare Kaon decays may well take the form of "facilities" where continual improvements are made as experience is gained in handling the backgrounds for a given mode. The full flux may well not be utilized in the initial stages, but should be available as improvements are made to counter problems as they appear.

WHAT LIMITS A $K \rightarrow \pi\nu\bar{\nu}$ EXPERIMENT?

R. E. Mischke
Meson Physics Division, Los Alamos National Laboratory
Los Alamos, NM 87545

ABSTRACT

Experimental limitations regarding the search for the decay $K \rightarrow \pi\nu\bar{\nu}$ are briefly reviewed.

Fourteen years ago the first limit[1] on the decay $K \rightarrow \pi\nu\bar{\nu}$ was published; it provided a test of neutral weak currents, complementary to those from limits on $K \rightarrow \mu\mu$ and $K \rightarrow ee$. Based on our present understanding of the number of neutrinos and quark masses and mixing angles, this decay is expected[2] to occur with a branching ratio between 4×10^{-10} and 3×10^{-9}. A branching ratio outside these limits would indicate the presence of new physics. It is also of interest to use this decay to look for new (unobservable) particles such as heavy neutrinos or supersymmetric particles.

Before discussing the factors that limit experiments studying this decay, it is useful to review the previous searches and the experiments planned for the near future. The general idea is to stop K mesons, identify a pion from the decay, and ensure that no other charged or neutral particles accompany the pion. The problems involve possible misidentification of the pion and possible escape of photons. (By definition any neutrals other than photons constitute an interesting decay mode.)

The first experimental search for $K \rightarrow \pi\nu\bar{\nu}$ was performed in a heavy-liquid bubble chamber.[1,3] It achieved an upper limit of 1.0×10^{-4} (90% C.L.) from a sample of 206,000 K^+ decays. Since then, counter techniques have been used to search for this decay. The next experiment[4] to set a limit for this decay mode used optical spark chambers, photographic recording of oscilloscope traces, and low beam rates. By looking for pions with energy greater than allowed from the decay $K \rightarrow \pi^+\pi^0$, it achieved a limit of 1.4×10^{-6}. A similar experiment[5] detected pions with energies below those from $K^+ \rightarrow \pi^+\pi^0$ and when combined with the limit above resulted in an improvement to 5.6×10^{-7}. The most recent attempt[6] used the same method as employed ten years earlier but used multiwire proportional chambers and a scintillation-counter telescope to reach the present upper limit of 1.4×10^{-7}. This experiment stopped a total of 1.5×10^{10} K^+ but the effective acceptance was only 1.6% due to the solid angle of the detector and kinematic cuts required to eliminate background from $K^+ \rightarrow \pi^+\pi^0$.

The next generation of experiments should establish that the decay exists. A third generation of experiments will be needed to gather sufficient statistics to study the decay in detail. Experiments planned for the near future have as a goal to reach the level of 10^{-10}. This appears feasible with present techniques and beam quality. One proposal[7] from CERN would use a large liquid argon detector with a stopping K^+ beam of 7.5 x 10^4/pulse. The advantage of a liquid-argon sphere is the large solid angle and the ability to detect the π^+ decay sequence and provide a photon veto in the same detector. The disadvantages are the need to locate a position sensitive detector at the center of the sphere and the need to separate Cherenkov and scintillation light from the argon for particle identification. Another proposal[8] in preparation for the AGS would use a scintillator stopping target surrounded by a cylindrical drift chamber, solenoidal magnetic field, and a 4π photon veto.

A protoproposal[9] to do this experiment at LAMPF II has been based on the plans being made for the AGS experiment. It estimates that with a stopping K^+ flux of 10^7/s, a trigger consisting of a single stopping K and a π^+ identified by range would give a trigger rate of ~10/s. The background rejection of other K^+ decays comes from the π^+ decay signal to reject $K^+ \rightarrow \mu\nu$ and $K^+ \rightarrow \mu\nu\gamma$ and from a pion energy cut and photon veto to reject $K^+ \rightarrow \pi^+\pi^0$. The rejection ratios are better than 10^{-13} for these modes.

It is very important to the success of these experiments to avoid pileup. The requirement of one and only one stopped K is satisfied by monitoring the incident beam to identify an incident K and then waiting a few nanoseconds for the decay signal. The identification of a pion requires a wait of additional tens of nanoseconds for the pion to decay. During this time it will be necessary to eliminate confusion from other incident or decaying particles. If we take 200 nanoseconds as a reasonable window for a single candidate this limits the incident flux to 5 x 10^6/s or less.

Present stopping K beams are more than an order of magnitude below this limit but even if more intense primary beams were available, the first step would be to improve beam quality rather than flux. For example, a typical π:K ratio is 5:1 after separation. This puts a substantial load on the singles capability of the detector. A higher primary proton beam flux would allow better separation reducing the pion contamination in the beam. Also, more primary flux would allow a smaller momentum bite for the K beam for a given flux. This would allow a thinner stopping target and would further reduce the pion interactions that could contribute to the detector singles rate. In the limit the pion flux disappears and the K flux is increased until either the singles rate from K decay is the maximum permitted by the detector or the limit of a "confusion free" time window is reached.

Assuming the detector singles rate is the limitation, further improvements can only come with improvements in detector technology. Similarly, to eliminate the need for a wide time window, improvements in the sophistication of the analysis of the experiment is required. Neither of these is a fundamental limitation and it is expected that technological development will keep up with the availability of more intense beams.

In summary, present accelerators have sufficient intensity to provide enough kaons for an experiment to observe the decay $K \to \pi\nu\bar{\nu}$ at the 10^{-10} level. A substantial increase in flux would make even the next generation of experiment significantly easier and it will be essential for experiments that hope to collect large statistics. Present detector technology is adequate for available fluxes; in the Kaon Factory era, the singles rates and confusion from random backgrounds are expected to limit the rate of progress of these experiments. It is not possible to predict in detail the limiting factors for an experiment with adequate flux. Experience shows that an improvement of one order of magnitude is good for a next experiment of this type. At each level one finds new problems that cannot be anticipated and must be solved. The ingenuity of the experimenters is always taxed to find a solution that optimizes the usage of the resources available.

REFERENCES

1. U. Camerini et al., Phys. Rev. Letters 23, 326 (1969).
2. J. Ellis and J.S. Hagelin, Nucl. Phys. B217, 189 (1983).
3. D. Ljung and D. Cline, Phys. Rev. D8, 1307 (1973).
4. J.H. Klems et al., Phys. Rev. D4, 66 (1971).
5. G.D. Cable et al., Phys. Rev. D8, 3807 (1973).
6. Y. Asano et al., Phys. Letters 107B, 159 (1981).
7. M. Ferro-luzzi et al., CERN Proposal No. CERN/PSCC 82-24.
8. L. Littenberg and W.C. Louis III, private communication.
9. W.C. Louis III, submission to LAMPF II physics proposal.

LEPTON MIXING AND HEAVY NEUTRINOS

J.-M. Frere
Randall Laboratory of Physics
University of Michigan, Ann Arbor, MI 48109

In view of the considerable interest displayed at this meeting in experiments looking for $K \to e\nu_x$, $K \to \mu\nu_x$, where ν_x is an hypothetical heavy neutrino mixing with the electronic and/or muonic neutrino, I think it may be appropriate to comment on a more general (although less sensitive) approach to such mixings.

Once we accept the possibility of having heavy neutrinos there is no a priori reason to assume that they are lighter than the K. Although K decays provide us with extremely strict bounds on the mixing of such neutrinos, they are necessarily only sensitive to a relatively small window of the possible mass spectrum.

More general bounds can be obtained by comparing neutral and charged-current data at low energy. The basic idea is very simple; if very heavy neutrinos are present and mix with ν_e and ν_μ they will "sterilize" part of the charged current (e.g. if $m_{\mu_x} > m_\mu$, the rate $\mu \to e\nu\nu$ will be decreased in proportion of the mixing). The departure from μ-e universality will not be observable if the mixing is small, or, more precisely, if the difference between ν_μ-ν_x and ν_e-ν_x mixing is small.[1] On the other hand, heavy neutrinos do not influence the neutral current or electric coupling of charged leptons. Therefore, a direct comparison between neutral and charged-current data should provide us with some information (bound) on the mixing with heavier neutrinos.

The issue is somewhat complicated by our relative ignorance of the exact value of M_W and M_z. How much can then be said from low energy experiments?

From the above, it should be clear that the determination of $\rho = \frac{M_W}{M_z \cos\theta}$ from ν_μ, ν_e scattering and the e-d experiment would be biased by neutrino mixing, in which case ρ should be replaced by a parameter $\rho' > \rho$. (Since the charged currents are suppressed by the mixing, the apparent value of M_W is increased.) The fact that in the standard model radiative corrections to the relation $\rho = 1$ are usually positive,[2] and that the presently observed value is close to unity,[3] already puts a limit of 6% on the mixing between ν_μ and <u>any</u> combination of heavy neutrinos.[4] More precise low energy measurements and/or a precise knowledge of the W mass should allow us to improve considerably on this.

REFERENCES

1. B.W. Lee and R.E. Shrock, Phys. Rev. D16, 3444 (1974).
2. M. Einhorn, D.R.T. Jones and M. Veltman, Nucl. Phys. B191, 146 (1981).
3. See for instance, M. Davier, in Proceedings, 21st International Conference on High Energy Physics, Paris, 26-31 July, 1982, page C3-474.
4. For more details, see J.-M. Frere, Nucl. Phys. B177, 389 (1981).

Theoretical Perspectives on Strange Physics

JOHN ELLIS*
Stanford Linear Accelerator Center
Stanford University, Stanford, California 94305

1. General Overview

Kaons are heavy enough to have an interesting range of decay modes available to them, and light enough to be produced in sufficient numbers to explore rare modes with satisfying statistics. Kaons and their decays have provided at least two major breakthroughs in our knowledge of fundamental physics. They have revealed to us CP violation,[1] and their lack of flavor-changing neutral interactions warned us to expect charm.[2] In addition, $K^0 - \bar{K}^0$ mixing has provided us with one of our most elegant and sensitive laboratories for testing quantum mechanics.[3] There is every reason to expect that future generations of kaon experiments with intense sources would add further to our knowledge of fundamental physics. This talk attempts to set future kaon experiments in a general theoretical context, and indicate how they may bear upon fundamental theoretical issues.

Figure 1 encapsulates some important trends in elementary particle physics. Two major philosophical approaches can be distinguished. The "onion-skin" philosophy emphasizes the search for more elementary constituents of matter as previous levels are shown to be composite. Many authors believe[4] that the present "elementary" particles such as quarks, leptons, possibly gauge bosons and maybe Higgs fields are in fact composites of underlying preons on a distance scale $\leq O(10^{-16})$ *cm*. The unification merchants, on the other hand, emphasize[5] the common origin and form of the different fundamental forces. It now seems established that the weak and electromagnetic interactions are at least partially unified in the Glashow-Salam-Weinberg (GSW) model,[6] and that the nuclear forces are complicated manifestations of an underlying gauge theory QCD[7] which is conceptually close to the GSW model. Together they constitute the "Standard Model" of elementary particles. Their family resemblance leads naturally to the hypothesis[8] of a grand unified theory of all gauge interactions. If this exists, its grand symmetry must be broken at a very high energy scale[9] within a few orders of magnitude of the Planck mass of $O(10^{19})$ GeV. At that scale there may be a final "superunification" with gravity. The provocative prefix "super" reflects my belief that such a final unification probably employs supersymmetry (SUSY) in some form.[10] It may well be that SUSY makes an earlier appearance on the stage of unification. As we will see later, technical difficulties with GUTs would be alleviated if SUSY was effectively restored at an energy scale as low as 1 TeV.

* Work supported by the Department of Energy, contract DE-AC03-76SF00515.

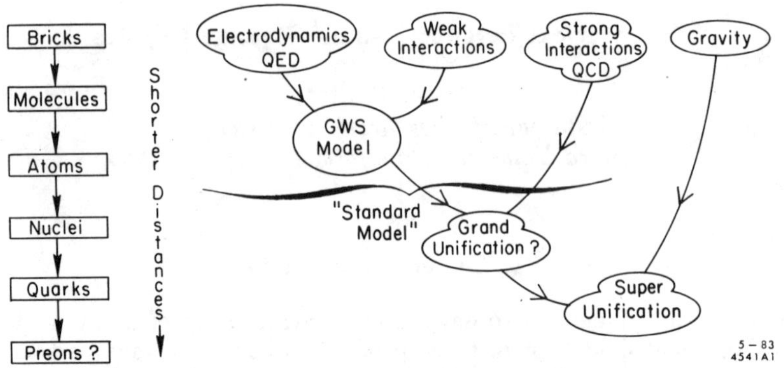

Fig. 1. A general overview of the "onion skin" and unification trends in elementary particle physics.

The different theoretical ideas introduced above are developed in greater detail in Sec. 2 of this talk. Then follows a survey of different experiments which would be done with an Intense Medium Energy Source of Strangeness, including rare K decays, probes of the nature of CP-violation, μ decays, hyperon decays and neutrino physics. Each experiment will be assessed for its interest as a test of the different theoretical ideas reviewed in Sec. 2. Section 3 concludes with a personal list of priorities for IMESS experiments. That terminates the physics content of this talk, and leaves us with two short sections of craziness. Section 4 discusses why and how quantum mechanics might be violated, and how one might test this in the $K^0 - \bar{K}^0$ system. Finally, Sec. 5 asks the unaskable question: how best should one proceed to explore strange physics in the future?

2. Survey of Different Theories

2.1 The Standard Model

By this name we denote the gauge theory based on the group $SU(3)$ (for the strong interactions) $\times SU(2) \times U(1)$ (for the weak and electromagnetic interactions), with just 3 (or perhaps $N > 3$) identical generations of quarks and leptons, and with spontaneous gauge symmetry breaking furnished by a single doublet of Higgs fields.[11] As mandated by the suppression of flavor-changing neutral interactions of kaons, the Standard Model incorporates the GIM mechanism[2] so that neutral currents conserve flavor at the tree level, and flavor-changing neutral interactions at the one-loop level are suppressed by $O(\alpha\, m_q^2/m_W^2)$. The GIM mechanism is an automatic consequence of any theory[12] in which all quarks of the same charge and helicity have the same weak isospin and get their masses from the same Higgs doublet. The Standard Model provides not only a qualitative but also a quantitative explanation of the magnitudes of flavor-changing neutral interactions. For example, the magnitude of the $\Delta S = 2$ $K^0 - \bar{K}^0$ mass mixing was used[13] to give an experimentally verified upper limit on the mass of the charmed quark, and the c quark contribution to the $\Delta S = 1$ $K^0 \to \mu^+\mu^-$ decay matrix element is small enough[13] to be compatible with experiment.[14] We will return later to these calculations in the 6-quark Kobayashi-Maskawa[15] extension of the original 4-quark GIM model.[2] They will provide us with useful constraints on the angles θ_i ($i = 1, 2, 3$) characterizing the charged weak interactions of quarks,[16] as well as on the mass of the t quark[17] and thereby connect with the phase δ which is the sole source of CP violation in the Standard Model. There is no room for weak mixing angles between leptons in the Standard Model as the neutrinos are supposed to be massless. This means that the numbers $L_{e,\mu,\tau}$ of the different Lepton families are absolutely conserved, so that $\mu \not\to e\gamma$, $K^0 \not\to \mu e$, etc. Even if the neutrino masses are non-zero, upper limits[18] on their values suppress $\Delta L \neq 0$ reactions to unobservably low rates unless there is some physics beyond the Standard Model.

2.2 More W's or Higgs?

It is natural to entertain the possibility of extending the weak gauge group of the Standard Model, perhaps by making it more symmetric: $SU(2) \times U(1)$ becomes $SU(2)_L \times SU(2)_R \times U(1)$ with parity broken spontaneously, or perhaps by making it more unified: $SU(2)_L \times U(1)$ becomes $SU(3)_L$ or? In the absence of any attractive unified weak interaction models the unifiers[5] have gone on to grand unification[8] at a scale $\geq O(10^{15})$ GeV.[9] Left-right symmetric models are still very much with us, as evidenced by discussions in the CP-violation section of this meeting.[19] They predict two more $W_R^\pm$ beyond the two $W_L^\pm$ already discovered, and two neutral Z^0 bosons. Left-right symmetric models also expect a right-handed neutrino field and non-zero neutrino masses. The right-handed neutrino can acquire an $SU(2)_L \times U(1)$ invariant Majorana mass and may well be rather heavy. There are constraints[20] on the masses of $W_R^\pm$ bosons and their mixing with $W_L^\pm$ bosons which are considerably more restrictive if the ν_R are lighter than a kaon.[22]

Why not have more Higgs bosons? They are needed in theories with larger gauge groups, and could easily be incorporated in the minimal $SU(2)_L \times U(1)$ theory. With two Higgs doublets one can implement[23] a global $U(1)$ symmetry which would solve the problem of strong CP-violation in QCD through the θ vacuum parameter.[24] SUSY theories actually require an even number of Higgs doublets in order to cancel out anomalies and give masses to all quarks and leptons.[25] Another motivation for multiple Higgses was to get an additional source of CP-violation[26] as discussed in the CP-violation session at this meeting.[19] However, in models with a $U(1)$ symmetry and in SUSY theories with two Higgs doublets, the quarks of charge $+\ 2/3$ get their masses from one Higgs doublet and the quarks of charge $-\ 1/3$ from another. Thus these models incorporate the GIM mechanism,[2] the neutral Higgs couplings conserve flavor, and there is no extra Higgs source[26] of CP-violation. In general, theories with two Higgs doublets contain two charged bosons $H^\pm$ and three neutrals including two scalars H^0 and $H^{0\prime}$ and one pseudoscalar a which becomes the light axion if a global $U(1)$ symmetry[23] is implemented.

2.3 Dynamical Symmetry Breaking

Many people regard Higgs fields as an unattractive wart on the face of gauge theory which they would prefer to burn out. One suggestion[27] is that Higgses are in fact composites of fermions bound together by some new "technicolor" gauge interaction which confines them within a range of $[\Lambda_{TC} = 0(1\,\text{TeV})]^{-1}$. The role of the spontaneous symmetry breaking previously associated with the vacuum expectation of elementary Higgs fields is now usurped by dynamical symmetry breaking associated with condensates $<0|\bar{F}F|0>$ of these "technifermions." This mechanism gives masses to the vector bosons in a very economical way, but requires epicycles in order to give masses to quarks and leptons. One proposed solution[28] was to introduce new "extended technicolor" (ETC) interactions as shown in Fig. 2.

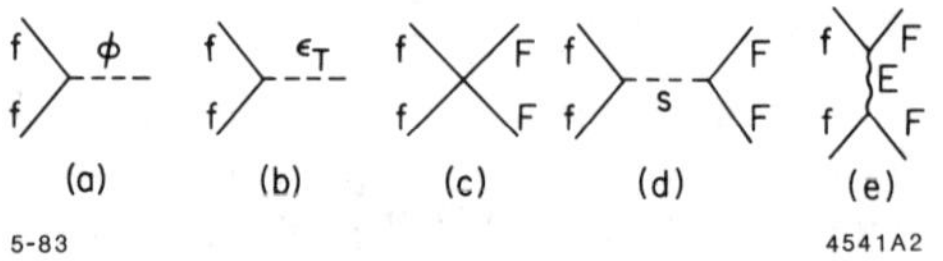

Fig. 2. (a) A Higgs–$f\bar{f}$ vertex metamorphoses into (b) a composite scalar $\epsilon_T - f\bar{f}$ vertex which requires (c) a four-fermion vertex that can be generated either by (d) scalar exchange or (e) vector ETC (E) exchange.

The conventional elementary Higgs–$\bar{q}q$ vertex is replaced by the composite Higgs–$\bar{q}q$ vertex which embodies a four-fermion $\bar{Q}Q\bar{q}q$ interaction generated by the exchange of massive ETC gauge bosons. One then has quark and lepton masses

$$m_{q,L} \approx \frac{\Lambda_{TC}^3}{m_{ETC}^2} \tag{1}$$

enabling the ETC boson masses to be estimated in terms of the known fermion mass spectrum. In a favored class[29] of ETC models there is one technigeneration (U, D, E, N) of techniquarks and leptons in parallel to the conventional generations (u, d, e, ν) etc. They are coupled to the conventional fermions by several different classes of ETC bosons, namely at least one per generation as indicated in Fig. 3a. In addition to these gauge bosons, a non-Abelian gauge theory of ETC must contain bosons coupling the different conventional generations to one another as seen in Fig. 3b. These "horizontal" ETC bosons have the same properties as the horizontal gauge bosons often postulated in the absence of an ETC motivation. However, in ETC theories the masses of these horizontal bosons are similar to those of the ETC bosons and can be estimated[30] using formula (1). There are other denizens of the technicolor zoo, some of whose masses are less tightly constrained. The theory contains[29] many "technipions" which are partners of the composite Higgs that escape being eaten by the $W^\pm$ and Z^0. They include color triplet bound states of techniquarks and technileptons called pseudoscalar leptoquarks P_{LQ} whose masses are expected[29] to be $O(150)$ GeV. There are also color singlet charged pseudoscalars $P^\pm$ which should[31] have masses $O(5$ *to* $14)$ GeV and have not been seen at PEP or PETRA,[32] to the embarrassment of technicolor theorists. There should also be even lighter neutral

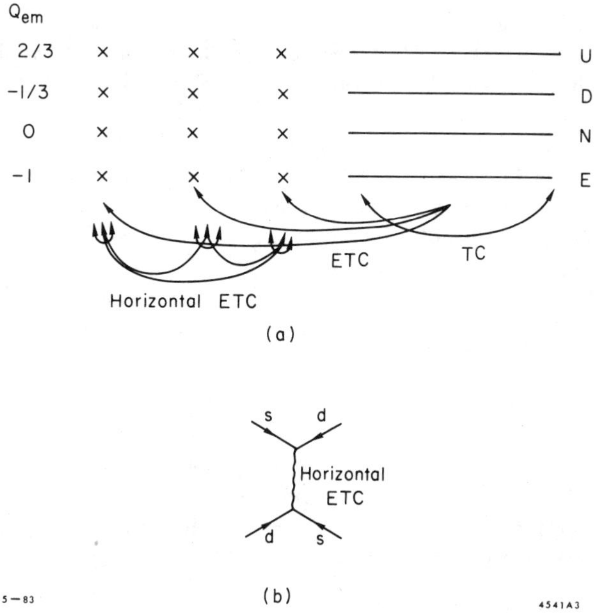

Fig. 3. (a) A sketch of the group structure of the simplest extended technicolor theories, and (b) a typical flavor-changing neutral interaction mediated by horizontal ETC boson exchange.

bosons $P^{0,3}$ whose masses can only[33] come from other interactions that we have not yet mentioned. One candidate[34] is vector leptoquark interactions of the Pati-Salam[35] type. These are expected[34,36] to yield

$$m_{P^{0,3}} = O\left(1\frac{1}{2}\right) \text{ GeV } \times \left(\frac{300 \text{ TeV}}{m_{PS}}\right) \times O\left(2^{0\pm1}\right) \tag{2}$$

and the non-observation of $K^{\pm} \to \pi^{\pm} P^{0,3}$ decay tells[36] us that

$$M_{P^{0,3}} \geq O(350) \text{ MeV} \tag{3}$$

which suggests via Eq. (2) an upper bound on m_{PS} of order 3000 TeV in such extended ETC theories. These vector leptoquarks will be met again in the discussion of rare K and Σ decays, along with the pseudoscalar leptoquarks.

Before leaving ETC theories, we should emphasize that models of the type described have severe problems with flavor-changing neutral interactions,[30] especially the magnitude of the $\Delta S = 2$ interaction responsible for $K^0 - \bar{K}^0$ mixing, and the absence so far of a $\Delta C = 2$ interaction leading to $D^0 - \bar{D}^0$ mixing. These problems can be traced[30,36,37] to the failure of ETC theories to satisfy the usual conditions[12] for natural flavor conservation. People have not abandoned hope of solving these problems.[38] If and when a full solution is found, it may well affect some of the order of magnitude estimates of rare K decays that we make later. However, these estimates do apply to the models[36,37] with partial solutions that do exist.

2.4 SUPERSYMMETRY

This is another response to the puzzles posed by Higgs fields. In order for the $W^{\pm}$ and Z^0 bosons to have required masses of $O(100)$ GeV, there must be at least some Higgs bosons with comparable masses. However, elementary scalar fields are believed to receive contributions to their masses $\delta m_H = 0(m_P \simeq 10^{19}$ GeV) when propagating (Fig. 4a) through the space-time foam that is believed to constitute the quantum gravitational vacuum.[39] More prosaically, they acquire $\delta m_H = 0(m_X \simeq 10^{15}$ GeV) from interactions (Fig. 4b) with other Higgses in the grand unified theory vacuum.[40]

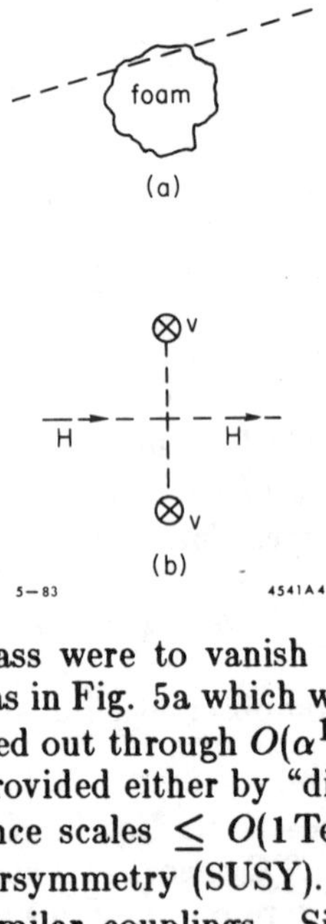

Fig. 4. A scalar field acquires large mass by propagating either through (a) space-time foam or (b) through the GUT vacuum.

Even if these contributions to the light Higgs mass were to vanish or cancel miraculously, there would be radiative corrections as in Fig. 5a which would give $\delta m_H^2 = 0(\alpha^n)(m_P^2$ or $m_X^2)$. These must be cancelled out through $O(\alpha^{12})$, which requires some quite powerful magic. This can be provided either by "dissolving" Higgses so that they become composite on distance scales $\leq O(1\,\mathrm{TeV}^{-1})$, as in technicolor theories,[27] or else by imposing supersymmetry (SUSY). In SUSY theories[10] there are bosons and fermions with similar couplings. Since their quantum loops have opposite signs as indicated in Fig. 5b, fermions and bosons tend to cancel so that

$$\begin{aligned} \delta m_H^2 &\approx \frac{g^2}{16\pi^2} \int^{\Lambda \sim m_X \text{ or } m_P} d^4k \frac{1}{k^2} \approx \frac{g^2}{16\pi^2}\Lambda^2 \\ &\to \frac{g^2}{16\pi^2}(m_B^2 - m_F^2) \ . \end{aligned} \tag{4}$$

In order for δm_H to be less than the required Higgs masses $O(100$ GeV) we see from Eq. (4) that one needs

$$(m_B^2 - m_F^2) \leq O(1 \text{ TeV}^2) \ . \tag{5}$$

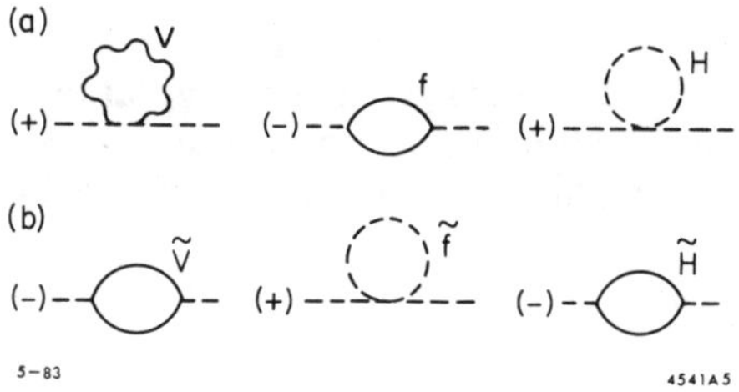

Fig. 5. (a) Diagrams renormalizing the scalar mass, and (b) the diagrams which almost cancel them in a SUSY theory, with signs indicated in parentheses.

Thus the unseen SUSY partners of known particles cannot be very heavy. The basic building blocks of the simplest SUSY theories[10] are supermultiplets containing pairs of particles differing in helicity by $\pm$ 1/2:

$$\begin{pmatrix} 1 \\ \frac{1}{2} \end{pmatrix} \begin{matrix} \text{gauge boson} \\ \text{gaugino} \end{matrix} \; ; \; \begin{pmatrix} \frac{1}{2} \\ 0 \end{pmatrix} \left. \begin{matrix} \text{quark } q \;, & \text{lepton } \ell \;, & \text{shiggs } \tilde{H} \\ \text{squark } \tilde{q} \;, & \text{slepton } \tilde{\ell} \;, & \text{Higgs } H \end{matrix} \right\} \tag{6}$$

Of this zoo of new particles, the lightest gaugino is likely to be the photino $\tilde{\gamma}$, while there may also be light neutral shiggs particles $\tilde{H}^0$.[41]

The same flavor-changing neutral interactions that were the Nemesis[30] of technicolor theories also impose strong constraints on SUSY theories.[42] The tree level couplings of all the new neutral particles must conserve flavor, and loop contributions to $\Delta F \neq 0$ interactions must be very small. The first requirement implies that the squark and slepton mass eigenstates must be spartners of pure quark and lepton mass eigenstates, and hence that the SUSY analogues of the Cabibbo-Kobayashi-Maskawa charged weak current mixing angles must be nearly identical with the familiar quark mixing angles. The suppression of loop diagrams requires a super-GIM mechanism, for example in the super-box diagram of Fig. 6 which is of order

$$G_F \frac{\alpha}{4\pi} \sin^2\theta_c \frac{m_c^2}{m_W^2} \tag{7}$$

only if

$$\frac{m_{\tilde{c}}^2 - m_{\tilde{u}}^2}{m_{\tilde{c}}^2 \text{ or } m_{\tilde{u}}^2} \simeq O\left(\frac{m_c^2}{m_W^2}\right) \leq O(10^{-3}) \; . \tag{8}$$

At first blush, these $\Delta F \neq 0$ neutral interaction conditions may seem difficult to satisfy, but in fact they emerge naturally in models with SUSY broken spontaneously. There all the squarks acquire universal $m_{\tilde{q}}^2$ of order (30 GeV)^2 or more, while all differences $\Delta m_{\tilde{q}}^2$ in squared masses are $O(\Delta m_q^2) \approx O(1)$ GeV^2 in the case of the first two generations. In these spontaneously broken SUSY theories super-loop contributions to $\Delta F \neq 0$ processes can be comparable[43] to the usual Standard Model contributions, and there may also be observable decays[44,49,46] into light SUSY particles such as the $\tilde{\gamma}$ and $\tilde{H}^0$, as we shall see later.

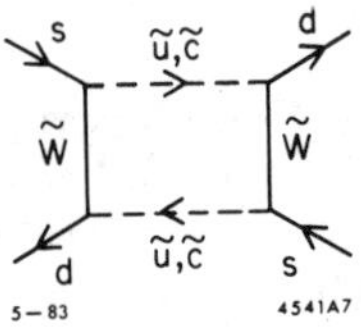

Fig. 6. A typical super-box diagram contributing to $\Delta F \neq 0$ neutral interactions in a SUSY theory which requires a super-GIM cancellation.

There are also SUSY models which have additional sources of symmetry violation beyond those in the Standard Model. For example, there exist possible additional sources of CP violation.[47] There is also the possibility of spontaneous lepton number violation due to vacuum expectation values for the spin-zero partners of neutrinos, whose effects are currently being investigated.[48] Indeed, we cannot even be sure that the photino and gluino couplings conserve flavor at the tree level. This is strongly suggested by the phenomenological constraints,[44,45,46] but it can be argued[49] that one should take a more agnostic phenomenological viewpoint.

2.5 Preons

It seems natural to suppose that the particles we currently regard as fundamental and elementary are in fact composite.[4] We have already removed so many layers of the onion — why not one more? Moreover, we now know such an untidy profusion of quark and lepton flavors that it is very appealing to seek a simpler description of nature with fewer fundamental elements. We have already toyed with the idea of composite Higgs fields, so perhaps quarks, leptons and even gauge bosons are composite also on a distance scale $O(\Lambda^{-1})$ which may be as large as $O(1\ \text{TeV}^{-1})$. Suppose for example that the only preons are fermionic, with quarks and leptons containing at least three, while bosons contain two preons. Then one can visualize the observed interactions as being due to exchange forces as in Fig. 7, with the dominant forces of longest range occurring in channels corresponding to the lightest bosonic bound states with masses in $\ll \Lambda$. This is similar to the way π exchange is important in nuclear interactions because $m_\pi \ll 1$ GeV. We would expect there to be additional forces in other channels corresponding to the exchanges of other bosons with masses $m = O(\Lambda)$ analogous to the ρ, ω and tensor meson exchanges of the conventional strong interactions.

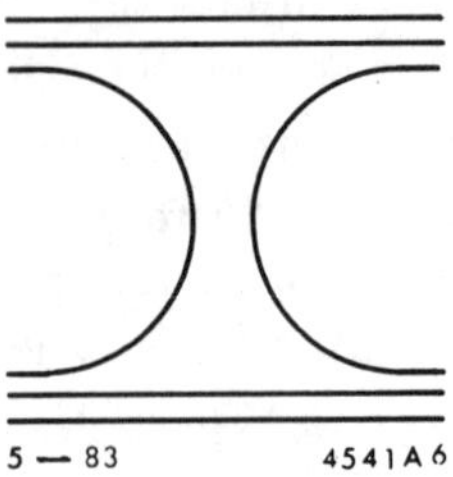

Fig. 7. An exchange diagram which yields a new effective interaction in a preon theory.

This analogy with the conventional strong interactions is too glib and glosses over many technical puzzles.[4] Why are the lightest bound states fermions with $m \ll O(\Lambda)$, whereas the fermionic bound states of QCD, the baryons, are heavy with masses $O(1\ \text{GeV})$? The underlying dynamics must obey consistency conditions[50] which are very difficult to satisfy. Why are there light bosons of spin 1, whereas the only light bosons in QCD have spin 0? It is very difficult to see how light gauge bosons would emerge unless the underlying dynamics already possessed the corresponding gauge invariance.

If we suspend our disbelief for a moment, we might expect that the exchanges of heavy bosons, or other dynamics on a scale of $O(1\ \text{TeV}^{-1})$ might generate all manner of novel interactions:

$$\frac{O(1)}{\Lambda^2}(\bar{q}q\bar{q}q\ ,\ \bar{q}q\bar{\ell}\ell\ ,\ qqq\ell\ ,\ \ldots) \tag{9a}$$

and more generally

$$\frac{O(1)}{\Lambda^{4-\frac{3n}{2}-m}} f^n B^m \tag{9b}$$

In general, we would not expect these new interactions to conserve conventional quantum numbers such as lepton number L, baryon number B, etc., since presumably different quark and lepton flavors share some preonic constituents in common, though some of the interactions (9) might be suppressed by some approximate chiral symmetry. The low-energy phenomenology of preon theories may in many ways resemble that of ETC theories, since observable new interactions of the form (9) probably involve the transformation of horizontal generation quantum numbers or the exchange of leptoquark quantum numbers.

3. Survey of Experimental Probes

3.1 Rare K Decays

There has been a lot of discussion[49] at this meeting of the whole gamut of rare K decays. Here I will only select a few possible experiments and concentrate on a subset of theories, treating composite models only cursorily and models with multiple gauge or Higgs bosons not at all. Each experiment will be discussed in turn for its interest within different frameworks, and the results are summarized in a table expressing my personal assessments.

3.1.1 $K_L^0 \to \mu e$ and $K^\pm \to \pi^\pm \mu e$ These two decays do not occur at all in the Standard Model, because it conserves separately both electron and muon lepton numbers L_e and L_μ. The same is also true in most SUSY models, though there is a possible escape route.[48] If any of the spin-zero sneutrino fields acquire a vacuum expectation value, the corresponding violation of lepton number would conceivably have repercussions in K decay, though the details are still being worked out.[48]

In contrast, $K_L^0 \to \mu e$ and $K^\pm \to \pi^\pm \mu e$ are mainstream possibilities in technicolor theories,[17] where there are many possible contributions to these decays. In the direct channel one can exchange a horizontal ETC gauge boson as in Fig. 8a. From the experimental upper limit[51]

$$\frac{\Gamma(K_L^0 \to \mu e)}{\Gamma(K^+ \to \mu^+ \nu_\mu)} < 0.63 \times 10^{-9} \tag{10}$$

one deduces[34] a lower limit on the squared mass of the corresponding gauge bosons:

$$m_{E_{d,s}}^2 \geq (3500\ ,\ 170)\ \text{TeV}^2 \tag{11}$$

which is exceedingly close to the estimates[30] obtained by using the formula (1). Therefore this mechanism should yield $K_L^0 \to \mu e$ decay at a rate very close to the experimental limit (10). Another contribution to $K_L^0 \to \mu e$ decay could come from the exchange of a Pati-Salam[35] vector leptoquark boson in the crossed channel as in Fig. 8b. In this case the limit (10) tells us[34] that

$$m_{PS} > O(300)\ \text{TeV} \tag{12}$$

whereas we deduced earlier from the nonobservation of $K^\pm \to \pi^\pm P^{0,3}$ decay that $m_{P^{0,3}} > O(350)$ MeV (3) and hence $m_{PS} < O(3000)$ TeV. Hence this mechanism should give rise to $K_L^0 \to \mu e$ decay at a rate within at most $O(10^{-4})$ of the present experimental upper limit (10). The exchanges of pseudoscalar bosons can also contribute to $K_L^0 \to \mu e$ decay. For example, the light pseudoscalars $P^{0,3}$ can be exchanged in the direct channel as in Fig. 8c, and the upper limit (10) tells us[30] that

$$m_{P^{0,3}}^2 > 2 \times 10^5 \theta^2\ \text{GeV}^2 \tag{13}$$

where the θ are some $\Delta F \neq 0$ mixing angles. Since the $P^{0,3}$ are expected on the basis of Eqs. (2) and (12) to weigh less than 3 GeV, the constraint (13) means that the mixing angles θ must be very small indeed. This possibility is not excluded within the ETC framework.[36] A contribution to $K_L^0 \to \mu e$ which is less easy to suppress is crossed channel pseudoscalar leptoquark (P_{LQ}) exchange as in Fig. 8d. The upper limit (10) tells us[34] that

$$m_{P_{LQ}} \geq O(150)\ \text{GeV} \tag{14}$$

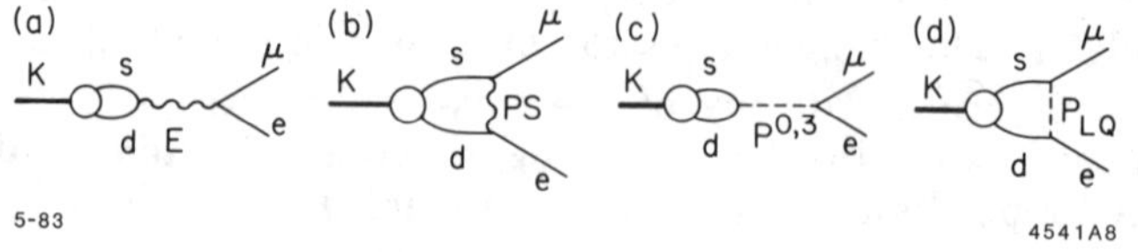

Fig. 8. Contributions to $K_L^0 \to \mu e$ decay from (a) direct channel HETC exchange, (b) cross channel Pati-Salam gauge boson exchange, (c) direct channel $P^{0,3}$ exchange and (d) crossed channel pseudoscalar leptoquark exchange.

which coincides with the best estimates[29,33] of the mass of the P_{LQ}. Therefore ETC theories would also expect $K_L^0 \to \mu e$ via this mechanism to occur very close to the present experimental upper limit (10). This contribution is difficult to avoid because there is no way of preventing technileptons and techniquarks from binding to form the P_{LQ}, its mass is simply related by scaling to the $\pi^+ - \pi^0$ mass difference, and the $s \to \mu$ and $d \to e$ transitions are not expected to be suppressed by any small angle factors.

The above contributions all have analogues in the decays $K^\pm \to \pi^\pm \mu e$. The only difference is that the present experimental limit[52]

$$\frac{\Gamma(K^\pm \to \pi^\pm \mu e)}{\Gamma(K^+ \to \pi^0 \mu^+ \nu_\mu)} < 1.5 \times 10^{-7} \tag{15}$$

is a weaker constraint than the $K_L^0 \to \mu e$ limit (10). For example, it means[30] that

$$m_{E_d}^2 > O(200)\ \mathrm{TeV}^2 \tag{16}$$

to be compared with the more stringent bound (11). However, the bound (15) is logically independent, since $K_L^0 \to \mu e$ proceeds via a pseudoscalar coupling to the quarks, whereas $K^+ \to \mu e$ proceeds via a vector or scalar coupling. Therefore in principle one decay could occur and not the other, though in practice models[37] predict comparable values for the ratios (10) and (15).

Searches for the decay $K_L^0 \to \mu e$ and also for $K^+ \to \pi^+ \mu e$ would be very interesting and topical from the point of view of technicolor theories of dynamical symmetry breaking.[29] These decays could also occur in preon models,[4] but the rates are much more difficult to pin down. However, as we see later they may be the best ways to test preon models in rare K decays.

3.1.2 $K^+ \to \pi^+ + \text{Higgs}$ A parenthetic revival of this decay is appropriate here since an experiment to look for $K^+ \to \pi^+ \mu e$ decay could also be a useful Higgs search experiment. In the Standard Model there is just one physical neutral Higgs which should weigh more than 10 GeV.[11] However, there could be a lighter neutral Higgs in models with multiple Higgs doublets, such as supersymmetric models for example. We expect[11] the following to be the dominant decay modes of light neutral Higgs bosons:

$$\begin{aligned}
2m_e < m_H < 2m_\mu &\ :\ H \to e^+e^- \text{ or } \gamma\gamma \\
2m_\mu < m_H < 2m_\pi &\ :\ H \to \mu^+\mu^- \\
2m_\pi < m_H < 2m_K &\ :\ H \to \mu^+\mu^- \text{ or } \pi^+\pi^-\,,\ \pi^0\pi^0 \\
2m_K < m_H < 2m_\tau &\ :\ H \to K\bar{K}\,(n\pi)\ .
\end{aligned} \tag{17}$$

Nuclear physics excludes[11] $m_H < O(15)$ MeV. There is a claim[53] that the absence of a Higgs peak in $\eta' \to \eta + \mu^+\mu^-$ excludes $m_H < O(400)$ MeV, but the experimental uper limit is not far below the calculation[54] of the $\eta' \to \eta + H$ decay rate which is not very reliable. The range $m_\pi < m_H < 2m_\mu$ is almost excluded by a search in the $K^+ \to \pi^+ + e^+e^-$ decay.[55] There is no information about lighter Higgses because of background problems, and heavier Higgses would have decayed into $\mu^+\mu^-$ and could not have been seen in this experiment. It would be interesting to perform sensitive searches for Higgses as spikes in $K^+ \to \pi^+ + (e^+e^-$ and $\mu^+\mu^-)$, which could be byproducts of $K^+ \to \pi^+\mu e$ experiments.

3.1.3 $K^+ \to \pi^+ +$ Nothing "Nothing" comes in many varieties of unobserved neutrals, including the following:

$K^+ \to \pi^+ +$ axion This is a two-body decay for which the experimental upper limit[57] is

$$\frac{\Gamma(K^+ \to \pi^+ + a)}{\Gamma(K^+ \to \text{all})} < 3.8 \times 10^{-8} \tag{18}$$

to be compared with a theoretical rate[58] which is generally $\geq O(10^{-6})$. It may be possible to arrange for a partial cancellation in the decay amplitude by a judicious choice of charged Higgs boson mass, but it seems very difficult to suppress the decay rate below the upper limit (18). This is just one of the many reasons why the conventional axion should be dead. However, its Frankenstein refuses to accept[59] the mortality of his creature: perhaps it is an undead zombie which still awaits the silver stake of another experiment to be driven through its heart?

$K^+ \to \pi^+ +$ familon Wilczek[60] has invented another light pseudoscalar for you, this time the exactly massless familon f, a Goldstone boson of a conjectured family or generation symmetry. He estimates[60] the branching ratio

$$B(K^+ \to \pi^+ + f) = O(10^{14})\left(\frac{1\ \text{GeV}}{F}\right)^2 \tag{19}$$

where F is an analogue of the pion decay constant f_π. Cosmology[61] tells us that $F \leq O(10^{12})$ GeV, whereas present experiments (18) tell us that $F \geq 5 \times 10^{10}$ GeV. This small gap could be closed by an experiment sensitive to a branching ratio of $O(10^{-10})$. Here is a theory which could be excluded by a forthcoming rare K decay experiment.

$K^+ \to \pi^+ + \sum_i \nu_i \bar{\nu}_i$ The experimental upper limit[57] on this three-body decay is

$$\frac{\Gamma(K^+ \to \pi^+ + \sum_i \nu_i \bar{\nu}_i)}{\Gamma(K^+ \to \text{all})} < 1.4 \times 10^{-7} \ . \tag{20}$$

The GIM loop diagrams[62] of the Standard Model give for each neutrino flavor i

$$B(K^+ \to \pi^+ \nu_i \bar{\nu}_i) = 0.6 \times 10^{-6}\, \frac{|\tilde{D}_i|^2}{U_{us}^2} \times \text{ QCD factor} \tag{21}$$

with

$$\tilde{D}_i = \sum_{j>1} U_{js}^* U_{jd} \bar{D}\left(\frac{m_{q_j}^2}{m_W^2} , \frac{m_{L_i}^2}{m_W^2}\right) . \tag{22}$$

In these formulae, the QCD correction factor is reasonably well known, the U_{jk} denote Cabibbo-Kobayashi-Maskawa mixing matrix elements, and the $\bar{D}_i$ are known[62] kinematic functions of the quark and lepton masses. As can be seen in Fig. 9, $\bar{D}$ vanishes when the mass of the lepton associated with ν_i is $O(1 \text{ to } 3) \times m_W$. The Cabibbo-Kobayashi-Maskawa mixing angle factors and the unknown t quark mass are constrained by other $\Delta F \neq 0$ amplitudes[63] such as $K_L^0 \to \mu^+\mu^-$ which gives an upper bound[17] on the $K^+ \to \pi^+ + \nu_e \bar{\nu}_e$ decay rate[44] as seen in Fig. 10. The known amount of $K^0 - \bar{K}^0$ mixing provides[44] a lower bound on the $K^+ \to \pi^+ + \nu_i \bar{\nu}_i$ (for light associated charged leptons, ℓ_i) which can also be seen in Fig. 10. Indicated explicitly in Fig. 10 is the uncertainty in the $K^0 - \bar{K}^0$ mixing constraint arising from our ignorance of the $\Delta S = 2$ operator matrix element $< \bar{K}^0 |(\bar{s}\, d)^2| K^0 >$, usually expressed as a coefficient R times the value obtained by inserting the vacuum intermediate state.

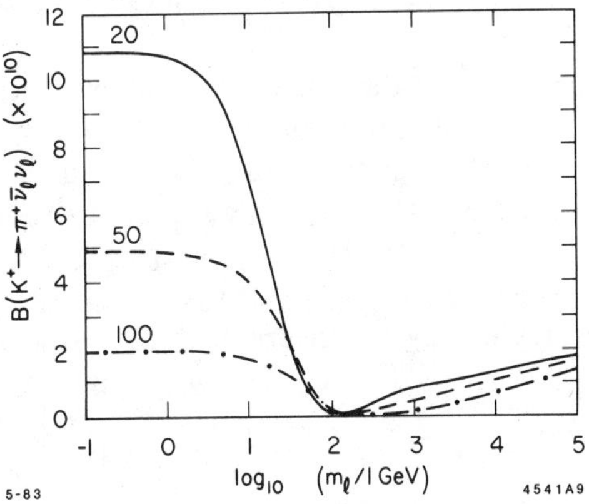

Fig. 9. Variation of the $K^+ \to \pi^+ \bar{\nu}_\ell \nu_\ell$ branching ratio with the mass of the lepton ℓ associated with ν_ℓ, for different values of the t quark mass in GeV.

There was a recent advance[64] on this point when it was realized that this matrix element would be related by $SU(3)$ and PCAC to the known $\Delta I = 3/2$ $K^+ \to \pi^+\pi^0$ decay amplitude, yielding $R = 0.33$ with perhaps a (30 to 60)% uncertainty indicated by the dashed lines in Fig. 10. Another potential uncertainty arises in the $K_L^0 \to \mu^+\mu^-$ constraint. In contrast to the $\Delta S = 2$ operator, it can get significant contributions from SUSY super-loop diagrams,[43,65] which could lift the bound for a given value of m_t. If we take conservatively the solid lines in Fig. 10 corresponding to $R = 1$ and no SUSY contribution to $K_L^0 \to \mu^+\mu^-$, we find they are only compatible for $m_t < 60$ GeV, the Buras[17] bound. We get[44] the absolute upper bound

$$\sum_i B(K^+ \to \pi^+ + \nu_i \bar{\nu}_i) < N_\nu \times 1.1 \times 10^{-9} \tag{23}$$

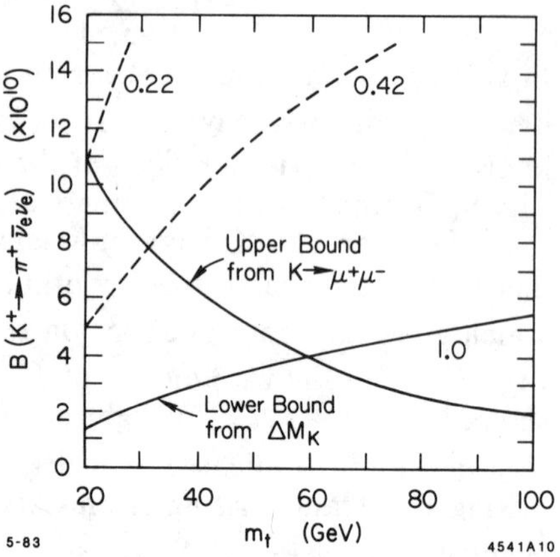

Fig. 10. Upper and lower bounds on the $K^+ \to \pi^+ \bar{\nu}_e \nu_e$ decay rate for different values of m_t. The numbers refer to the matrix element factor R.

on the decay rate and a lower bound[44]

$$\sum_i B(K^+ \to \pi^+ + \nu_i \bar{\nu}_i) > N_\nu^{light} \times 1.4 \times 10^{-10} \tag{24}$$

where N_ν^{light} denotes the number of neutrinos whose associated leptons are light. The numbers (23) and (24) neglect the contributions of the heavier quarks to the GIM loop diagrams. Because of the $K_L^0 \to \mu^+\mu^-$ and $K^0 - \bar{K}^0$ mixing constraints, they can only slightly increase the upper bound on the $K^+ \to \pi^+ + \nu_i \bar{\nu}_i$ decay rate.[44] However, the lower bound can be destroyed by cancellations between the t quark and the 8th quark, though this cannot[44] happen if the 8th quark mixing angles are in the hierarchial ratio

$$\left|\frac{U_{j8}}{U_{jt}}\right|^2 \approx \frac{m_t}{m_8} \tag{25}$$

If we ignore all this uncertainty, the bound (24) can be used to establish an upper bound on the number of neutrinos whose associated charged leptons do not have masses too close to the W mass as seen in Fig. 11. The curve was plotted using the conservative value $R = 1$ for the $\Delta S = 2$ matrix element.

Our conclusion is that the decay $K^+ \to \pi^+ +$ unobserved neutrals is very interesting in the context of the Standard Model. If the branching ratio is significantly larger than the upper limit of 3.3×10^{-9} (23) this would be a signal for new physics: perhaps more than three neutrinos or .....? Paradoxically, a branching ratio significantly lower than 4.2×10^{-10} (24) could also be a signal for new physics, perhaps a cancellation between third and fourth generation quarks, or .....?

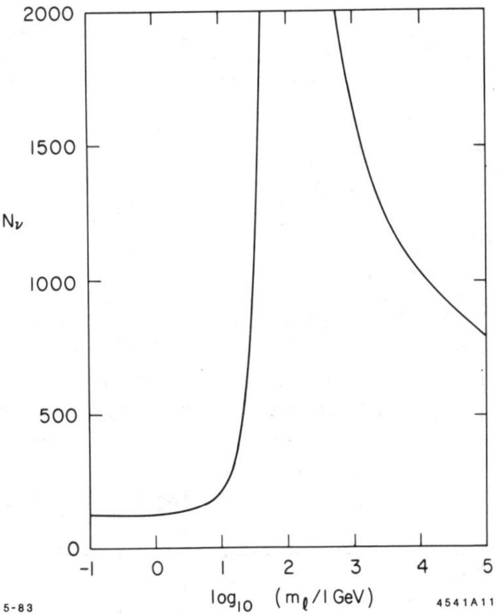

Fig. 11. The present upper bound on the number of neutrinos from $K^+ \to \pi^+ + \bar{\nu}_\ell \nu_\ell$ decay[22] as a function of the mass m_ℓ of the charged lepton associated with ν_ℓ.

$K^+ \to \pi^+ + \nu_i \bar{\nu}_{j \neq i}$ So far we have concentrated on processes where the outgoing neutrino and antineutrino have the same flavor. ETC theories and preon models could also give neutrinos and antineutrinos of different flavors, by mechanisms similar to those by which they give $K^+ \to \pi^+ \mu e$. Unfortunately, since these decays are experimentally indistinguishable from $K^+ \to \pi^+ + \nu_i \bar{\nu}_i$, they will not be identifiable unless their branching ratios are above a few times 10^{-9}, which seems unlikely in view of the $K_L^0 \to \mu e$ bound (10).

$K^+ \to \pi^+ +$ photinos $\tilde{\gamma}$ or shiggses $\tilde{H}^0$ The light neutral particles expected in the SUSY theories of Sec. 2.4 could also be pair-produced in K^+ decays, either at the tree level,[45,49] or more plausibly (?) via super-loop diagrams analogous to those for decays into $\nu_i \bar{\nu}_i$. We expect[44,46] decays into gravitino pairs to be negligible, and we expect

$$B(K^+ \to \pi^+ + \tilde{\gamma} + \text{ gravitino}) \leq 2B(K^+ \to \pi^+ + \tilde{\gamma} + \tilde{\gamma}) \qquad (26)$$

and it is probably much smaller unless the scale of SUSY breaking is surprisingly small. The photino pair-production rate depends on the unknown spectrum of SUSY particles, but in general[44,46]

$$B(K^+ \to \pi^+ + \tilde{\gamma} + \tilde{\gamma}) \leq O\left(\frac{1}{10}\right) B(K^+ \to \pi^+ + \nu_i \bar{\nu}_i) \qquad (27)$$

so that it is unlikely to push the rate for $K^+ \to \pi^+ +$ nothing up substantially. However, in some theories the lightest SUSY particles may be neutral shiggses $\tilde{H}^0$, and they could[44] be pair-produced with branching ratios comparable with those for a single neutrino flavor:

$$\frac{B(K^+ \to \pi^+ + \tilde{H}^0 \tilde{H}^0)}{B(K^+ \to \pi^+ + \nu_e \bar{\nu}_e)} = \frac{\left(-7 - 5\, ln \frac{m_t^2}{m_W^2}\right) v_1^2 + \left(-4 + 2\, ln \frac{m_t^2}{m_W^2}\right) v_2^2}{\left(-1 - 3\, ln \frac{m_t^2}{m_W^2}\right) (v_1^2 + v_2^2)} \qquad (28)$$

where v_1 and v_2 are the vacuum expectation values of the two Higgs doublets in SUSY theories. We expect $v_1 \geq v_2$, and if $v_1 \ll v_2$ the ratio (28) becomes 0.88 for $m_t = 20$ GeV. We conclude that the decay $K^+ \to \pi^+ +$ nothing can also be interesting for SUSY theories.

3.2 Comparisons

My personal assessments of the interest of different rare K decay experiments from the points of view of different theories are shown in Table 1. The crosses indicate frameworks where a given process does not occur. This does not necessarily mean that the experiment is not interesting from the point of view of that theory. As Sherlock Holmes has observed, the best clue is often the dog that does not bark!

Table 1. Testing Theories in Rare K Decays

Experiments \ Theories		Standard Model	Technicolor	Super-Symmetry	Preon Models
$K_L^0 \to \mu e$		×	√√√√	√ ?	√
$K^+ \to \pi^+ \mu e$		×	√√	√ ?	√
	light pseudo scalar	×	did not occur	×	×
$K^+ \to \pi^+$ + nothing	$\nu \bar{\nu}$	√√√√√	√	×	√
	SUSY particles	×	×	√√√	×

The number of checks in each box correspond to the amount of interest I personally have in confirming the non-zero prediction of the corresponding theory for that experiment. Do they correspond to the number of years of an experimentalist's life that it might be worth devoting to that test of the theory? The question marks denote cases where the theory is not yet completely clear.

Since Table 1 attaches considerable significance to $K^+ \to \pi^+ +$ nothing experiments, it is worthwhile to make comparisons with other nothing production experiments, in which the existence of nothing is inferred by tagging the event with a pion, photon, $\pi\pi$ or whatever.

$\pi^0 \to$ Nothing Clearly one way of tagging this is in $K^+ \to \pi^+ + (\pi^0 \to \text{nothing})$, and it has been discussed as a way of producing pairs of massive photinos[66] or neutrinos. There was some discussion at the Workshop of making and tagging the π^0 in another way, perhaps by $\pi^+ + d \to p + p + (\pi^0 \to \text{nothing})$, but it was concluded that background problems made this uncompetitive with K^+ decay.

$J/\psi \to$ Nothing This is in principle a reliable way to count neutrinos, using the $\psi' \to J/\psi + \pi^+\pi^-$ decay to tag $J/\psi \to$ nothing, but it is not very sensitive. From the expected[67] decay rate

$$\frac{B(J/\psi \to \bar{\nu}\,\nu)}{B(J/\psi \to e^+e^-)} \simeq 2 \times 10^{-8} N_\nu \Big[m_{J/\psi} \ (\text{GeV}) \Big]^2 \tag{29}$$

and the experimental upper limit[68] of 1/10 on this ratio we can infer that

$$N_\nu \lesssim 5 \times 10^5 \ . \tag{30}$$

The cosmologists[69] with their upper limit of three or four neutrinos should be laughing at us particle physicists! The constraints on N_ν from $K^+ \to \pi^+ +$ nothing[44] and from the late stages of stellar evolution[70] are much more stringent than (30), even though somewhat uncertain. As for SUSY particles, the rate for $J/\psi \to \tilde{\gamma}\tilde{\gamma}$ depends sensitively on the unknown $\tilde{c}$ squark masses,[71] and, in fact, requires a difference between the masses of partners of the left- and right-handed $\tilde{c}$ squarks.

$\Upsilon \to$ Nothing Many of the same remarks apply, except that the greater mass of the Υ means that one has better sensitivity than (29), but experimentalists have not yet got around to quoting an upper limit on $\Upsilon \to$ nothing. If they could establish that this was $\leq O(1/10) B(\Upsilon \to e^+e^-)$ then one would get a limit: $N_\nu \leq O(5000)$, which begins to be comparable with present limits from $K^+ \to \pi^+ +$ nothing decay.

$e^+e^- \to$ Nothing This can be tagged[72] by a bremsstrahlung γ, and there is currently a proposal[73] to look for these at PEP with a sensitivity corresponding to $N_\nu \leq O(10)$ at a center-of-mass energy of $\sqrt{s} = 29$ GeV. This search is also sensitive to photinos[74] if they weigh less than about 10 GeV and $M_{\tilde{e}} \leq O(50)$ GeV.

$Z^0 \to$ Nothing This is of course the neutrino counting experiment par excellence, and it is easy[75] to gain a sensitivity to $\Delta N_\nu \ll 1$ in the reaction $e^+e^- \to Z^0 + \gamma$. In contrast, $Z^0 \not\to \tilde{\gamma}\tilde{\gamma}$ at the tree level.

The reactions listed above were arranged by order of increasing energy, and hence sensitivity to heavier "nothing" particles. Since different experiments see different mass ranges as well as having differing sensitivities to different species of "nothing" particle, a complete understanding of the spectrum of quasi-stable neutral particles will require performing all the experiments. For example, if one found an apparent "N_ν"> 3 in both $Z^0 \to$ nothing and $e^+e^- \to$ nothing at $\sqrt{s} = 29$ GeV, would $K^+ \to \pi^+ +$ nothing tell us $N_\nu = 3$, in which case we would suspect the existence of novel neutrals, probably with masses larger than 150 MeV? Or if we were satisfied from Z^0 and $e^+e^- \to$ nothing that $N_\nu = 3$, we could use the rate for $K^+ \to \pi^+ +$ nothing as a probe of Standard Model dynamics. Either way, the decay $K^+ \to \pi^+ +$ nothing has an important role to play.

The same probably cannot be said for the decay $\mu \to e +$ nothing which can be probed at a π or K factory.[76] One difficulty is that the dominant decay of the μ is $\mu \to e +$ (nothing $= \nu\bar{\nu}$), so that there is a big background. One does not usually expect L_e and L_μ to be violated in spontaneously broken SUSY theories, so $\mu \not\to e + \tilde{\gamma} + \tilde{\gamma}$, whereas the analogous $K^+ \to \pi^+ + \tilde{\gamma} + \tilde{\gamma}$ decay can occur at $O(1/10)$ of the $K^+ \to \pi^+ + \nu + \bar{\nu}$ rate. For the same reason one does not expect $\mu \to e +$ axion decay, whereas the absence of $K^+ \to \pi^+ +$ axion has already been used to dispute the very existence of an axion. However, the two-body decay $\mu \to e +$ familon could be interesting:

$$B(\mu \to e + f) \approx O(10^{12})\left(\frac{1 \text{ GeV}}{F}\right)^2 \tag{31}$$

to be compared with the analogous $K^+ \to \pi^+ + f$ decay rate (19). For this reason a continued search for the two-body $\mu \to e +$ nothing decay down to a limit of order 10^{-12} may still be valuable.

3.3 Muon and Hyperon Decays

Now that we have broached the subject of $\Delta L_\mu \neq 0$ processes, it is worthwhile to make a systematic comparison of one with another, and also with the rare K decays that were the subject of Sec. 3.1.1. Table 2 is a compilation of limits from various rare processes[77] on the possible masses of certain species of exotic beasts. It is adapted from a table developed by Shanker.[78] It is always difficult to achieve a consensus on the couplings to be assumed for imaginary particles, and I have made somewhat different assumptions than Shanker.[78] These do not affect the ratios of different limits on the same particle from different experiments so much as they affect the absolute values of the limits, especially for theories with many Higgs doublets. Shanker[78] assumed couplings to leptons of order $g_2\, m_\tau/m_W$ and to quarks of order $g_2\, m_b/m_W$, whereas I have chosen values three orders of magnitude smaller. These values correspond more closely to the masses of the light quarks and leptons involved in the processes we are considering. If they appear too small, recall that there are probably mixing angle factors which enter when we consider reactions that violate L_e and/or L_μ, and probably also when one changes quark generation: $s \to d$. For this reason it is difficult to interpret directly the order-of-magnitude limit on the Higgs mass coming from processes violating L_e and L_μ. However, it probably is reasonable to interpret Table 2 as indicating that among such processes, multiple Higgs theories are most sorely tested by anomalous muon conversion $\mu Z \to eZ$. This need not be the case for other theories of L_e and L_μ violation, which may well be more severely tested by $\mu \to e\gamma$ or $\mu \to e\bar{e}e$ searches.

Table 2. Comparison of Mass Limits From Rare Processes

Process	Multiple Higgs	Pseudoscalar Leptoquark	Vector Leptoquark	Experimental Limit
$B(\mu \to e\gamma)$	0.2	–	–	1.9×10^{-10}
$B(\mu \to e\bar{e}e)$	0.4	–	–	1.9×10^{-9}
$B(\mu Z \to eZ)$	11.0	$0.15 \times \theta$	$60 \times \theta$	7×10^{-11}
$B(K_L^0 \to \mu e)$	7.0	0.18	93	2×10^{-9}
$B(K_L^0 \to \mu\mu)$	4.7	$0.12 \times \theta$	$62 \times \theta$	9×10^{-9}
$B(K_L^0 \to ee)$	7.0	$0.18 \times \theta$	$95 \times \theta$	2×10^{-9}
$B(K^+ \to \pi^+\mu e)$	0.7	0.01	3.5	7×10^{-9}
Δm_K	150	–	–	$\approx 3.5 \times 10^{-15}$ (GeV)
	mass limits in GeV	mass limits in TeV	mass limits in TeV	

Table 2 features my personal guesses as to possible mixing angle factors in leptoquark interactions which would arise if one makes the normal identifications of lepton and quark generations: $s \leftrightarrow \mu$, $d \leftrightarrow e$. We see again that the best limits on leptoquark masses seem to come from the $K_L^0 \to \mu e$ decay limit discussed in Sec. 3.1.1, with $K^+ \to \pi^+\mu e$ providing less stringent limits as we expected. We also see that the decays $K_L^0 \to \mu\mu$ and ee give less interesting limits because of mixing angle factors. Since the decay $K_L^0 \to \mu\mu$ has already been observed at a rate close to the unitarity limit, after $K_L^0 \to \mu e$ the next most interesting of these leptonic K decays to look for may be $K_L^0 \to ee$.

Another decay which gives access to similar physics is $\Sigma \to p\mu e$. Unfortunately, it has been estimated[77] that the upper limit on $K \to \mu e$ decay already suggests that

$$B(\Sigma \to p\mu e) \leq O(10^{-12}) \qquad (32)$$

By comparison, the most stringent upper limits over rare Σ decays are around 10^{-6}, which prompted one experimentalist at this meeting to describe the range (32) as "an awful long ways to go." K decays seem more sensitive probes of L_e and L_μ violating physics.

One point to be recalled in looking at Table 2 is that as the sensitivity to a rare decay branching ratio B is increased, the sensitivity to heavy boson masses only increases as

$$m_{heavy} \propto B^{-\frac{1}{4}} \qquad (33)$$

Thus an order of magnitude increase in sensitivity corresponds to a change in the limit on m_{heavy} by less than a factor of two. Moreover, this change is often small compared with other uncertainties in the calculation, such as those in the values of coupling constants and of matrix elements. Progress will not be rapid, nor will its interpretation be unambiguous.

3.4 CP-VIOLATION

The discovery[1] of CP-violation in the neutral K^0 system has so far been one of the greatest contributions by kaons to our physical knowledge. Unfortunately, despite its cosmic significance[79] this original manifestation of CP-violation in the $K^0-\bar{K}^0$ mass matrix is still the only CP-violating phenomenon observed experimentally. Table 3 presents a list of interesting CP-violating observables together with the corresponding predictions of the theoretical frameworks introduced in Sec. 2.

We see that four of the observables appear in K decays, while two involve hyperon decays. First we have the fundamental K_1-K_2 mass mixing parameter ϵ. Its value can be fitted but not predicted in the Standard Model where

$$\epsilon \;\propto\; s_2 s_3 \, sin\delta \text{ for small } \theta_2, \theta_3 \quad . \tag{34}$$

The observed value of $\epsilon = O(10^{-3})$ can be understood very naturally if s_2 and s_3 are both considerably smaller than $sin\theta_c$, as would be the case if the b quark lifetime turns out to be $O(10^{-12})$ seconds. In models with additional gauge bosons and left-right symmetric models in particular the value of ϵ is related to the masses of the heavier gauge bosons. In multiple Higgs models the value of ϵ is related to the spectra and couplings of the Higgs bosons. Technicolor models tend to predict magnitudes of the CP-conserving real part of the $\Delta S = 2$ $K_1 - K_2$ mass mixing which are much too large, and give no understanding why ϵ should be so small.[30] The last column in Table 3 introduces the non-perturbative QCD CP-violating θ vacuum angle,[24] which cannot contribute significantly to ϵ because of the severe upper limit of $O(10^{-9})$ on θ coming from the non-observation[80] of a neutron electric dipole moment. Next we turn to model predictions for intrinsic CP violation ϵ' in the $K \to 2\pi$ decay amplitude and the ratio ϵ'/ϵ. Figure 12 depicts a Penguin diagram, which plays a crucial role in calculations of this quantity. As reviewed by Wolfenstein,[19] this meeting has witnessed considerable progress in the elucidation of the Standard Model[81] and multi-Higgs model[82] predictions. The Standard Model predicts[81] a definite sign (positive) as well as a lower bound on the magnitude. Multiple Higgs models also predict[82] a definite sign (negative) and a magnitude much larger than the Standard Model bound. These models are on the borderline of experimental exclusion, but may not yet have crossed it. Left-right symmetric gauge models are unfortunately rather less specific in their predictions for ϵ'/ϵ. The naive technicolor models seem[83] to predict too large a value for ϵ'/ϵ, but this defect might also be rectified if one could cure[38] the other flavor-changing neutral interaction problems[30] of such theories.

Table 3. CP Violating Observables and Model Predictions

	Standard Model	Multiple W	Multiple H	Technicolor	Super-symmetry	θ_{QCD}
ϵ	fit	$\leftrightarrow m_{W_R}$	$\leftrightarrow m_H$	too big?		0
ϵ'/ϵ	$\geq 2 \times 10^{-3}$	$\sim 10^{-3}$	$\leq -2 \times 10^{-2}$	large ?		0
$\eta_{+-0} - \eta_{+-}$	$O(\epsilon')$	$O(10^{-5})$	?	large ?	as for Standard Model	0
P_n^μ in $K \to \pi\mu v$	0	0	$\leq 5 \times 10^{-3}$	0		0
$\frac{B(\Lambda\to p)-B(\bar\Lambda\to\bar p)}{B(\Lambda\to p)+B(\bar\Lambda\to\bar p)}$	$\sim \frac{\epsilon'}{5}$	?	?	?		0
$\frac{\alpha(\Lambda\to p)+\alpha(\bar\Lambda\to\bar p)}{\alpha(\Lambda\to p)-\alpha(\bar\Lambda\to\bar p)}$	$\sim 3\epsilon'$	?	?	?		0
$d_n(e-cm)$	$< 10^{-30}$	$\sim 3 \times 10^{-27}$	$\sim 5 \times 10^{-26}$	$\sim 10^{-25}$	$\lesssim 10^{-25}$ (also for electron)	$3 \times 10^{-16}\theta$
$D^0 - \bar D^0$	very small	?	large ?	large ?	as for Standard Model	0
$B^0 - \bar B^0$	unobservable	?	large ?	large ?		0
Universe	too small	fit - unless CPX spontaneously	fit - unless CPX spontaneously	0	possible	$\Rightarrow$ large θ ? ?

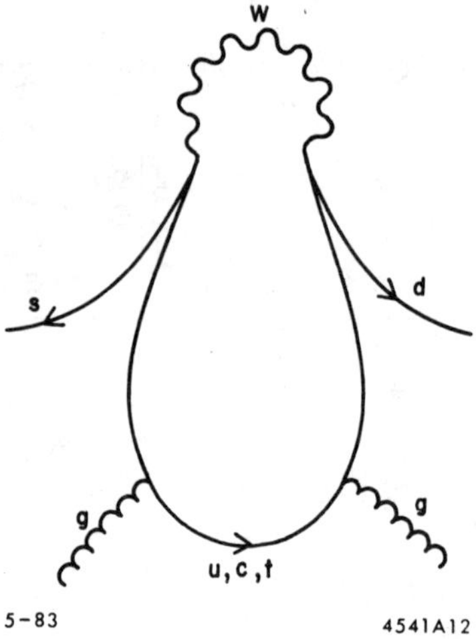

Fig. 12. A penguin diagram.

The next row[19] features predictions of the different models for intrinsic CP-violation in the $K \to 3\pi$ decay amplitude. It is expected to be the same order of magnitude as ϵ' in the Standard Model, and also very small in a left-right symmetric gauge model, whereas the multiple-Higgs model prediction has not been developed. The muon transverse polarization P_n^{μ} is expected to be non-zero only in multiple-Higgs models, but is not expected to be very large. For completeness, two CP-violating observables in hyperon decay are listed, together with the Standard Model predictions for them which are both $O(\epsilon')$. The left-right gauge and multiple-Higgs model predictions have not yet been developed.

We now turn to CP-violating observables beyond the ambit of the strange physicist. The neutron electric dipole moment d_n is expected[84] to be unobservably small in the Standard Model, but can be close to the present experimental limit[80] of 6×10^{-25} $e-cm$ in left-right gauge models,[85] multiple-Higgs models,[86] and in technicolor theories.[87] This is the only observable for which the QCD θ vacuum parameter is likely to play a role, and the estimate[88] $d_n \simeq 3 \times 10^{-16} \theta$ $e-cm$ tells us that

$$\theta \leq 2 \times 10^{-9} \tag{35}$$

which is why its effects are not observable elsewhere. Observation of a neutron electric dipole moment in the near future need not exclude the Standard Model of CP-violation. Its prediction could have been augmented by a contribution from θ close to the limit (35) which would not show up in any other phenomenological situation. The electron could also have an observable[89] electric dipole moment, which can be comparable with that of the neutron in some SUSY models.[90]

It is natural to ask whether CP-violation could appear in the $D^0 - \bar{D}^0$ or $B^0 - \bar{B}^0$ systems in ways analogous to its manifestation in the $K^0 - \bar{K}^0$ system. According to the Standard Model, ϵ should be small in the D^0 system but could be large in the B^0 system. This depends on the values of unknown Kobayashi-Maskawa angles, but unfortunately $B^0 - \bar{B}^0$ mixing is expected to be suppressed in domains of the angles where ϵ is large,[91] so it is unlikely that CP-violation could actually be observed in the B^0 system if the Standard Model is correct.

Finally we come to the Universe. It is included here because of the idea[79] that the observed baryon asymmetry in the Universe may have originated from CP- and B-violation in GUT reactions[92] when the Universe was $O(10^{-35})$ seconds old. This qualitative mechanism is not strong[93] enough in the Standard Model to produce the observed baryon-to-photon ratio of a few times 10^{-10}, but could be fit in more complicated models containing more Higgs multiplets and/or more gauge bosons. However, the connection with low energy physics is not clear, since the extra baryosynthetic structure need only appear at the GUT scale, and may not be light enough to show up in present-day accelerator experiments. Also, it is important to note that in many of these models with additional low energy structure CP is violated spontaneously, in which case no significant baryon asymmetry can be generated. For this reason technicolor theories[87] are not baryosynthetic. The QCD θ parameter also does not contribute directly to the baryon asymmetry,[94] although one can argue that most theories which generate enough baryons and do not possess some additional symmetry such as a Peccei-Quinn[23] $U(1)$ or SUSY will also have a value of θ close to the limit (35) and suggest a neutron electric dipole moment close to the present experimental limit.[95] Since cosmological CP-violation may be the reason we exist at this meeting, it provides a motivation for constructing a suitable extention of the Standard Model.

3.5 Neutrino Physics

The death of a kaon is often the birth of a neutrino which can be used for high intensity and precision neutrino physics, as was discussed by a working group[96] at this meeting. One can imagine detailed measurements of $\nu_\mu e$, $\bar{\nu}_\mu e$ and $\nu_e e$ scattering which enable one to measure $d\sigma/dy$ as well as σ. In this way one might be able to measure $sin^2\theta_W$ with a sensitivity comparable to that obtainable from experiments near the Z^0 peak. The comparison between low energy and high energy measurements is a crucial check to the radiative corrections[97] whose calculability was the prime motivation for spontaneously broken gauge theories. To get some idea of the precision required, let us recall that

$$\delta m_{Z^0} \approx 140 \text{ GeV} \times \delta(sin^2\theta_W) \tag{36}$$

so that a determination of m_{Z^0} with a precision of about 300 MeV would fix $sin^2\theta_W$ with an error of ± 0.002. It is not clear that one would gain from a more accurate value of m_{Z^0}, because of uncertainties in the one loop correction due to strongly interacting particles and because of higher order radiative corrections to the Z^0 mass. Compare this precision with the shift[98] in the apparent value of $sin^2\theta_W$ due to radiative corrections, which is $O(0.012)$. In this context, a "precision" low energy determination of $sin^2\theta_W$ should mean an error $\ll 0.01$ and preferably $O(0.002)$.

As discussed by Shrock[49] at this meeting, another interesting class of low energy neutrino experiments involves searches[99] for massive neutrinos ν_H. One may look for their production in $\pi \to \mu\nu_H$, $e\nu_H$ and $K \to \mu\nu_H$; $e\nu_H$ decays via anomalous bumps in the lepton energy spectrum, in which cases one is sensitive to $|U_{\mu H}|^2$ or $|U_{eH}|^2$ respectively, where the $U_{\ell H}$ are mixing matrix elements, as seen in Fig. 13. Another possibility is to search for ν_H decays, as was mentioned at this meeting by Ferro-Luzzi.[100]. One can look either for $\nu_H \to \nu_\ell e^+ e^-$ or for $\nu_H \to \nu_\ell \gamma\gamma$, and is generally sensitive to the product $|U_{eH} U_{eH}|$ or $|U_{eH} U_{\mu H}|$. Figure 13 is taken from a proposal[100] for such a decay experiment at CERN, and we see that very large ranges of mixing matrix elements are accessible to such an experiment. This range could be further improved using a more intense source, just as the μ or e spectrum bump-hunting experiments could also improve in sensitivity.

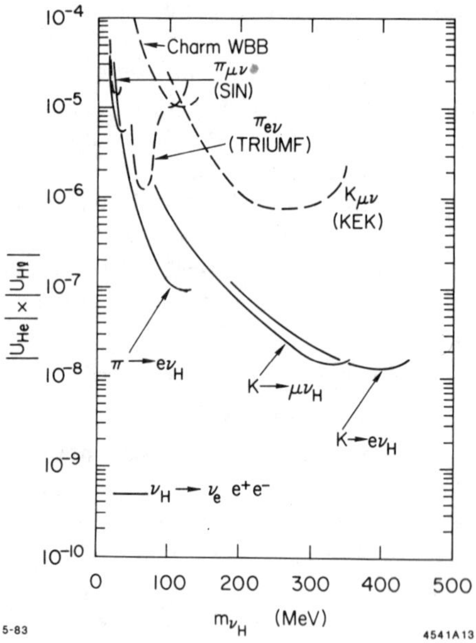

Fig. 13. Actual and potential bounds on hevy neutrino couplings as functions of their masses.

3.6 A Personal Priority List

It is always dangerous to put something like this into print, as it is likely to be over-simplified and over-interpreted. However, several participants have urged me to produce such a distillation, which I do not mind doing as long as everyone realizes it is just a personal opinion, and there are at least as many different opinions as there are theorists at this meeting.

Top of my priorities would be a measurement of the CP-violating parameter ϵ'/ϵ, because of the light it would cast on the mechanism(s) of CP-violation. Second would be $K^+ \to \pi^+ +$ nothing experiments, because this is the last great frontier of $\Delta S \neq 0$ neutral interactions which will tell us a lot about the Standard Model as well as expose us to new physics. Next I come to the more speculative searches for new interactions, at the top of which list I would place $K_L^0 \to \mu e$. This seems to be more sensitive and topical than the other searches such as $K^\pm \to \pi^\pm \mu e$, $\Sigma \to p\mu e$, and $\mu A \to eA$, $\mu \to e\gamma$ and $\mu \to e\bar{e}e$. At the bottom of my list come precision neutrino experiments, in part because of my doubts whether the errors could be reduced sufficiently to be really exciting. Then there are many other experiments which get an "incomplete."

4. Beyond Physics

The discussion of serious physics has terminated with the previous section. Now I would like to mention to you briefly a little philosophical speculation you may care to entertain: can one observe a violation of quantum mechanics?

It has been noticed[101] that black holes correspond to mixed thermal states with finite entropy associated to the existence of an event horizon across which information can be lost. Hawking[102] has further observed that quantum effects cause black holes to radiate particles with a mixed thermal spectrum. These discoveries may not be of purely philosophical or astrophysical significance in view of the old idea[103] that spacetime may have a foamy structure at short distances, with $O(1)$ mini black hole or other topological structure (instanton?) in every Planck volume. Thus we may imagine that event horizons constantly appear and disappear, and wonder whether this would have any implication for the purity of quantum mechanical wave functions. Hawking and collaborators[39] have performed calculations suggesting that indeed gravitational instantons may cause initially pure states to evolve with time into mixed states, and Hawking[104] has argued that the conventional laws of quantum mechanics should be modified as a consequence.

He entertains[104] the possibility that an initial density matrix ρ_{-D}^{C} may yield a final density matrix ρ_{+B}^{A}:

$$\rho_{+B}^{A} = \not{S}_{BC}^{A\ \ D}\, \rho_{-D}^{C} \tag{37}$$

where the superscattering operator $\not{S}_{BC}^{A\ \ D}$ does not factorize into the product of an S matrix and its adjoint:

$$\not{S}_{BC}^{A\ \ D} \neq S_C^A S_B^{\dagger\ D} \tag{38}$$

as in conventional quantum mechanics. If this is indeed the case, one might expect[105] a non-standard equation for the time evolution of the density matrix:

$$\frac{d\rho}{dt} = \not{H}(\rho) \tag{39}$$

where $\not{H}$ is a general hermitian linear operator which does not take the standard form:

$$\not{H} = i[\rho, H] + \delta\not{H} \ :\ \delta\not{H} \neq 0 \tag{40}$$

it is easy to check that pure states may evolve into mixed states if the $\not{S}$ operator does not factor as in Eq. (38), or if $\not{H}$ takes the non-canonical form (40). We would expect[106] the conventional rules of quantum mechanics to apply on time scales $\delta t \ll |\delta\not{H}|^{-1}$, but expect that if

$$\delta t \geq |\delta\not{H}|^{-1} \tag{41}$$

then violations of quantum mechanics may be observable. This general expectation[106] is indeed borne out by calculations in simple examples[105] of modified quantum mechanical systems.

For example, one might expect that an initially pure K^0 state produced in a hadron-hadron collision could evolve into a mixed state which is approximately a K_L but with a small K_S admixture. This K_S component can come from a term $\delta\, \not{H}$ in (40) which produces a slow 'decay' of the K_L into the K_S, appearing as a continuous 'regeneration' of K_S in vacuo (precisely what one would expect if the definite phase relationship between the K^0 and $\bar{K}^0$ components of a K_L beam were to become mixed in some way). This contributes to the downstream 2π yield in a way which is distinguishable from the usual CP impurity of the K_L, but this distinction requires a precise comparison of the CP parameters $|\eta_{+-}| \simeq |\epsilon|$, ϕ_{+-}, and $\delta \simeq 2Re\epsilon$. Based on existing data[18] this comparison gives

$$|\delta\, \not{H}| \leq 2 \times 10^{-21}\,\text{GeV} \quad . \tag{42}$$

A constraint on $|\delta\, \not{H}|$ for neutrons of similar magnitude can be deduced[105,106] from the success of long baseline neutron interferometry experiments. Since neutral kaons are at the cutting edge of this issue, it might be worthwhile to consider how one might devise somewhat more sensitive tests of quantum mechanics in the $K^0 - \bar{K}^0$ system.

5. A Question

Figure 14 is an artist's impression of the recent past history of elementary particle physics and its cousins which also features some possible extrapolations into the future. The horizontal lines depict fields of study which do not get more fundamental as time elapses. The diagonal line depicts the direction of elementary particle physics, including a few landmarks passed in the past as well as some possible landmarks of the future. As the diagonal and horizontal lines diverge, new fields of study open up between them. For example, the recent achievement of a consensus Standard Model of elementary particle physics bids fair to act as a node where a new line branches off horizontally. This may meet a new diagonal line branching off from traditional nuclear physics.

The old pion factories were mainly motivated by nuclear physics but have turned out to give some help in elucidating the Standard Model. Kaon factories would surely be useful for Standard Model physics. We can ask ourselves whether they will have significant impact on the particle physics of the 1990's. Assuming that they do, we should also ask our political masters whether we are playing in a zero-sum game. Would any significant fraction of the cost of a kaon factory be charged against the elementary particle physics account? If so, we particle physics chauvinists have to wonder whether kaon factories would be a cost-effective means of pursuing our discipline.

The "weak" lobby gathered at this meeting is agreed that strange physics is great physics. Furthermore, there seems to be little transatlantic competition since we learn that there is to be no beam for charged kaon physics at CERN in the foreseeable future. One has the impression that a kind of Yalta philosophy may be at work, according to which CERN concentrates on $\bar{p}$ physics and leaves strange physics to accelerators in the United States. Can the variety of strange physics we feel necessary be done "on the cheap"?

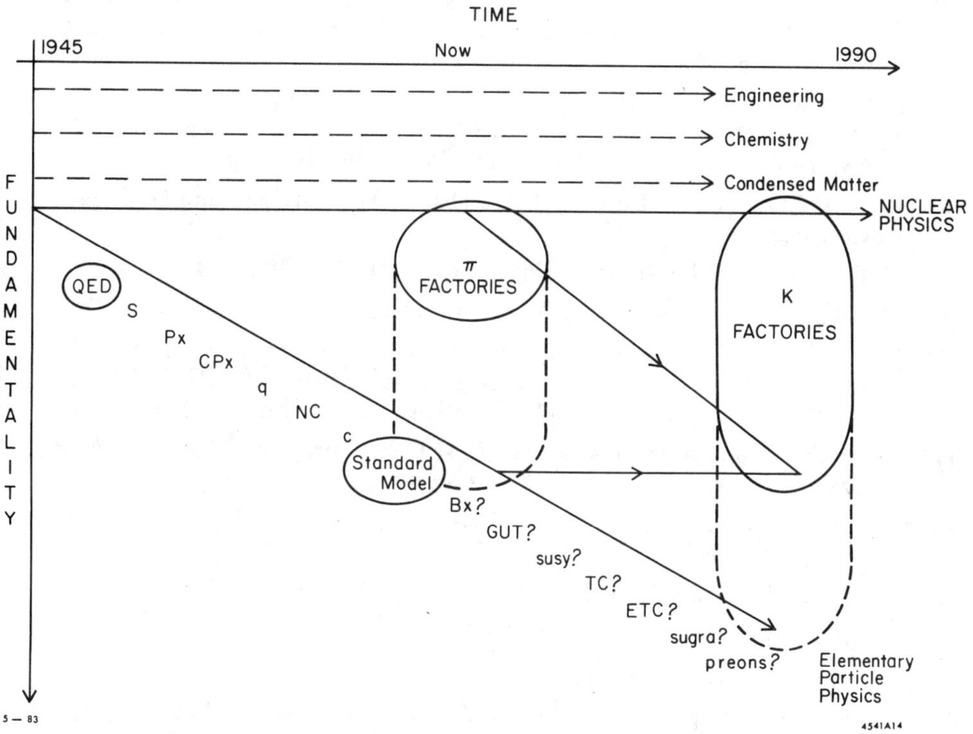

Fig. 14. A sketch of the past and possible future evolution of elementary particle physics and related disciplines.

Acknowledgements

It is a pleasure to thank T. Goldman, H. E. Haber and J. Primack for their kind hospitality, and to thank S. J. Brodsky, M. K. Gaillard, J. S. Hagelin, C. Heusch and C. Sachrajda for useful comments and discussions.

References

1. J. H. Christensen, J. W. Cronin, V. L. Fitch and R. Turlay, Phys. Rev. Lett. 13, 138 (1964).
2. S. L. Glashow, J. Iliopoulos and L. Maiani, Phys. Rev. D 2, 1285 (1970).
3. R. P. Feynman, The Feynman Lectures in Physics, Vol. III, p. 11-12 (Addison-Wesley, Reading PA, 1963).

4. For a review and references see M. E. Peskin, Proc. of the 10th Int. Symp. on Lepton and Photon Interactions at High Energies, ed., W. Pfeil (Univ. Bonn, 1981), p. 880.

5. For a review and references, see P. Langacker, Proc. of the 10th Int. Symp. on Lepton and Photon Interactions at High Energies, ed., W. Pfeil (Univ. Bonn, 1981), p. 823.

6. S. L. Glashow, Nucl. Phys. 22, 579 (1961); S. Weinberg, Phys. Rev. Lett. 19, 1264 (1967); A. Salam, Proc. 8th Nobel Symp., Stockholm 1968, ed., N. Svartholm (Almqvist and Wiksells, Stockholm, 1968), p. 367.

7. H. D. Politzer, Phys. Rep. 14C, 129 (1974); W. Marciano and H. Pagels, Phys. Rep. 36C, 137 (1978).

8. H. Georgi and S. L. Glashow, Phys. Rev. Lett. 32, 438 (1974).

9. H. Georgi, H. R. Quinn and S. Weinberg, Phys. Rev. Lett. 33, 451 (1974).

10. For reviews and references, see P. Fayet and S. Ferrara, Phys. Rep. 32C, 249 (1977); P. Fayet, Ecole Normale Supérieure preprint LPTENS 82/28 (1982), talk presented at Int. Conf. on High Energy Physics, Paris 1982.

11. For reviews and references, see J. Ellis, CERN preprint TH-3174/LAPP TH-48 (1981); M. K. Gaillard, these proceedings.

12. S. L. Glashow and S. Weinberg, Phys. Rev. D 15, 1958 (1977); E. A. Paschos, Phys. Rev. D 15, 1966 (1977).

13. M. K. Gaillard and B. W. Lee, Phys. Rev. D 10, 897 (1974).

14. M. Shochet et al., Phys. Rev. D 19, 1965 (1979).

15. M. Kobayashi and T. Maskawa, Prog. Theor. Phys. 49, 652 (1973).

16. For a review and references, see L.-L. Chau, BNL preprint (1982) to appear in Phys. Rep. C.

17. A. J. Buras, Phys. Rev. Lett. 46, 1354 (1981).

18. Particle Data Group, Phys. Lett. 111B, 1 (1982).

19. L. Wolfenstein and D. Chang, these proceedings.

20. T. G. Rizzo and G. Senjanović, Phys. Rev. Lett. 46, 1315 (1981); Phys. Rev. D 24, 704 (1981) and D 25, 235 (1982).

21. H. Abramowicz et al., CERN preprint EP/82-210 (1982).

22. Y. Asano et al., Phys. Lett. 104B, 84 (1981).

23. R. D. Peccei and H. R. Quinn, Phys. Rev. Lett. 38, 1440 (1977) and Phys. Rev. D 16, 1791 (1977).

24. G. 't Hooft, Phys. Rev. Lett. 37, 8 (1976) and Phys. Rev. D 14, 3432 (1976); R. Jackiw and C. Rebbi, Phys. Rev. Lett. 37, 172 (1976); C. G. Callan, R. F. Dashen and D. J. Gross, Phys. Lett. 63B, 334 (1976).

25. S. Dimopoulos and H. Georgi, Nucl. Phys. B193, 150 (1981). N. Sakai, Zeit. für Phys. C11, 153 (1982).

26. S. Weinberg, Phys. Rev. Lett. 37, 657 (1976).

27. S. Weinberg, Phys. Rev. D 13, 974 (1976) and Phys. Rev. D 19, 1277 (1979); L. Susskind, Phys. Rev. D 20, 2619 (1979).

28. E. Eichten and K. D. Lane, Phys. Lett. 90B, 125 (1980); S. Dimopoulos and L. Susskind, Nucl. Phys. B155, 237 (1979).

29. S. Dimopoulos, Nucl. Phys. B168, 69 (1980); M. E. Peskin, Nucl. Phys. B175, 197 (1980); J. P. Preskill, Nucl. Phys. B177, 21 (1981).

30. S. Dimopoulos and J. Ellis, Nucl. Phys. B182, 505 (1981).

31. S. Dimopoulos, Ref. 29; S. Chadha and M. E. Peskin, Nucl. Phys. B185 and Nucl. Phys. B187, 541 (1981); V. Baluni, Univ. of Michigan preprint UM HE 82-27 (1983).

32. M. Althoff et al., Phys. Lett. 122B, 95 (1983) and references therein.

33. R. Binétruy, S. Chadha and M. E. Peskin, Phys. Lett. 107B, 425 (1981) and Nucl. Phys. B207, 505 (1982).

34. S. Dimopoulos, S. Raby and G. Kane, Nucl. Phys. B182, 77 (1981); S. Dimopoulos, S. Raby and P. Sikivie, Nucl. Phys. B182, 449 (1981).

35. J. C. Pati and A. Salam, Phys. Rev. Lett. 31, 661 (1973); Phys. Rev. D 8, 1240 (1973); Phys. Rev. D 10 275 (1974).

36. J. Ellis, M. K. Gaillard, D. V. Nanopoulos and P. Sikivie, Nucl. Phys. B182, 529 (1982); J. Ellis, D. V. Nanopoulos and P. Sikivie, Phys. Lett. 101B, 387 (1981).

37. J. Ellis and P. Sikivie, Phys. Lett. 104B, 141 (1981).

38. R. Holdom, Phys. Rev. D 24, 157 (1981); A. Masiero, E. Papantonopoulos and T. Yanagida, Phys. Lett. 115B, 229 (1982); A. J. Buras, S. Dawson and A. N. Schellekens, Phys. Rev. D 27, 1171 (1983); A. J. Buras and T. Yanagida, Phys. Lett. 121B, 316 (1983); S. Dimopoulos, H. Georgi and S. Raby, Harvard preprint HUTP 83/A002 (1983).

39. S. Hawking, D. N. Page and C. N. Pope, Phys. Lett. 86B, 175 (1979) and Nucl. Phys. B170 (FS1), 283 (1980).

40. E. Gildener and S. Weinberg, Phys. Rev. D 13, 3333 (1976); E. Gildener, Phys. Rev. D 14, 1667 (1976); A. J. Buras, J. Ellis, M. K. Gaillard and D. V. Nanopoulos, Nucl. Phys. B135, 66 (1978).

41. J. Ellis, J. S. Hagelin, D. V. Nanopoulos and M. Srednicki, SLAC-PUB-3094 (1983).

42. J. Ellis and D. V. Nanopoulos, Phys. Lett. 110B, 44 (1982); B. A. Campbell, University of Toronto preprint (1981); M. Suzuki, U.C. Berkekey preprint UCB-PTH-82/8 (1982); O. Shanker, Nucl. Phys. B204, 375 (1982).

43. T. Inami and C. S. Lim, Prog. Theor. Phys. 65, 297 (1981).

44. J. Ellis and J. S. Hagelin, Nucl. Phys. B217, 189 (1983).

45. R. Shrock, Proc. 1982 DPF Summer Study on Elementary Particle Physics and Future Facilities, ed., R. Donaldson, R. Gustafson and F. Paige (APS, 1982), p. 291; see also M. Suzuki, Ref. 42.

46. M. K. Gaillard, Y.-C. Kao, I.-H. Lee and M. Suzuki, Phys. Lett. 123B, 241 (1983).

47. J. Ellis, S. Ferrara and D. V. Nanopoulos, Phys. Lett. 114B, 231 (1982); J. Ellis, D. V. Nanopoulos and K. Tamvakis, Phys. Lett. 121B, 123 (1983); J. Polchinski and M. B. Wise, Harvard preprint HUTP-83/A016 (1983); F. del Aguila, M. B. Gavela, J. A. Grifols and A. Méndez, Harvard preprint HUTP-83/A017 (1983).

48. L. Hall and J.-M. Frére, private communications (1983).

49. G. L. Kane and R. E. Shrock, these proceedings.

50. G. 't Hooft, Recent Developments in Gauge Theories, Cargése 1979, ed., G. 't Hooft, C. Itzykson, A. Jaffe, H. Lehmann, P. K. Mitter, I. M. Singer, R. Stora (Plenum Press, N. Y., 1980).

51. A. Clark et al., Phys. Rev. Lett. 26, 1667 (1971).

52. A. M. Diamant-Berger et al., Phys. Lett. 62B, 485 (1976).

53. R. I. Dzhelyadin et al., Phys. Lett. 105B, 239 (1981).

54. A. I. Vainshtein et al., Usp. Fiz. Nauk 131, 537 (1980).

55. J. Ellis, M. K. Gaillard and D. V. Nanopoulos, Nucl. Phys. B106, 292 (1976).

56. A. M. Diamant-Berger and R. Turlay, private communication (1976).

57. Y. Asano et al., Phys. Lett. 107B, 159 (1981).

58. T. Goldman and C. M. Hoffman, Phys. Rev. Lett. 40, 220 (1978); J.-M. Frére, M. B. Gavela and J. Vermaseren, Phys. Lett. 103B, 129 (1981).

59. H. Faissner, W. Heinrigs, A. Preussger and D. Samm, Aachen preprint PITHA 82/24 (1982), and references therein.

60. F. Wilczek, Phys. Rev. Lett. 49, 1549 (1982) and these proceedings.

61. J. Preskill, M. B. Wise and F. Wilczek, Phys. Lett. 120B, 127 (1982); L. Abbott and P. Sikivie, Phys. Lett. 120B, 133 (1982); M. Dine and W. Fischler, Phys. Lett. 120B, 137 (1982).

62. T. Inami and C. S. Lim, Nucl. Phys. B207, 533 (1982).

63. R. Shrock and M. B. Voloshin, Phys. Lett. 87B, 375 (1979).

64. J. F. Donoghue, E. Golowich and B. R. Holstein, Phys. Lett. 119B, 412 (1982). For a recent upper bound on the $\Delta S = 2$ matrix element, see B. Guberina, B. Machet and E. de Rafael, Marseille preprint CPT-83/P.1495 (1983).

65. A. B. Lahanas and D. V. Nanopoulos, private communication (1983).

66. M. K. Gaillard et al., Ref. 46. The analogous method of searching for massive neutrinos is independent of mixing angles, unlike those discussed in Sec. 3.5.

67. J. Rich and D. R. Winn, Phys. Rev. D 14, 1283 (1976); see also J. Ellis, Ref. 11.

68. M. Breidenbach, private communication quoted by P. Fayet, Proc. Europhysics Study Conf. on the Fundamental Particle Interactions, eds., S. Ferrara, J. Ellis and P. van Nieuwenhuizen (Plenum Press, N. Y., 1980), p. 587. It would be nice to have a published reference to which to refer.

69. K. Olive, D. N. Schramm, G. Steigman, M. S. Turner and J. Yang, Astrophys. J. 246, 557 (1981); J. Yang, M. S. Turner, G. Steigman, D. N. Schramm and K. Olive, to be published (1983).

70. J. Ellis and K. Olive, CERN preprint TH-3323 (1982).

71. J. Ellis and S. Rudaz, SLAC-PUB-3101 (1983).

72. E. Ma and J. Okada, Phys. Rev. Lett. 41, 287 (1978) and Phys. Rev. D 18, 4219 (1978); K.J.F. Gaemers, R. Gastmans and F. M. Renard, Phys. Rev. D 19, 1605 (1979).

73. D. Burke, et al., PEP Proposal 021 (1983).

74. P. Fayet, Phys. Lett. 117B, 460 (1982); J. Ellis and J. S. Hagelin, Phys. Lett. 122B, 303 (1983).

75. G. Barbiellini, B. Richter and J. Siegrist, Phys. Lett. 106B, 414 (1981).

76. This account is based on discussions at the meeting with T.-P. Cheng and L.-F. Li.

77. See also R. N. Cahn and H. Harari, Nucl. Phys. B176, 135 (1980); G. L. Kane and R. Thun, Phys. Lett. 94B, 513 (1980).

78. O. Shanker, Nucl. Phys. B206, 253 (1982).

79. A. D. Sakharov, Zh. Eksp. Teor. Fiz. Pis'ma Red 5, 32 (1967).

80. I. S. Altarev, et al., Phys. Lett. 102B, 13 (1981).

81. J. S. Hagelin and F. J. Gilman, these proceedings; F. J. Gilman and J. S. Hagelin, SLAC-PUB-3087 (1983).

82. N. G. Deshpande, talk at this meeting; A. I. Sanda, Phys. Rev. D 23, 2647 (1981); N. G. Deshpande, Phys. Rev. D 23, 2654 (1981); J. F. Donoghue, J. S. Hagelin and B. R. Holstein, Phys. Rev. D 25, 195 (1982); J. S. Hagelin, Phys. Lett. 117B, 441 (1983).

83. M. E. Peskin, private communication.

84. J. Ellis and M. K. Gaillard, Nucl. Phys. B150, 141 (1979); E. P. Shabalin, I.T.E.P. preprint 65-1982 (1982) and references therein; M. B. Gavela et al., Phys. Lett. 109B, 83, 215 (1982).

85. D. Chang, Nucl. Phys. B214, 435 (1983).

86. N. G. Deshpande, Ref. 82, and to be published (1983).

87. E. Eichten, K. D. Lane and J. Preskill, Phys. Rev. Lett. 45, 225 (1980).

88. V. Baluni, Phys. Rev. D 19, 2227 (1979); R. J. Crewther, P. DiVecchia, G. Veneziano and E. Witten, Phys. Lett. 88B, 123 (1979) and E 92B, 487 (1980).

89. M. B. Gavela and H. Georgi, Phys. Lett. 119B, 141 (1982).

90. M. B. Gavela, these proceedings.

91. J. S. Hagelin, Phys. Rev. D 20, 2893 (1979) and Nucl. Phys. B193, 123 (1981); E. Franco, M. Lusignoli and A. Pugliese, Nucl. Phys. B194, 403 (1981).

92. M. Yoshimura, Phys. Rev. Lett. 41, 381 (1978) and Phys. Lett. 88B, 294 (1979); A. Yu. Ignatiev, N. V. Krasnikov, V. A. Kuzmin and A. N. Tavkhelidze, Phys. Lett. 76B, 436 (1978); S. Dimopoulos and L. Susskind, Phys. Rev. D 18, 4500 (1978) and Phys. Lett. 81B, 416 (1979); D. Toussaint, S. B. Treiman, F. Wilczek and A. Zee, Phys. Rev. D 19, 1036 (1979); J. Ellis, M. K. Gaillard and D. V. Nanopoulos, Phys. Lett. 80B, 360 (1979); S. Weinberg, Phys. Rev. Lett. 42, 850 (1979).

93. J. Ellis, M. K. Gaillard and D. V. Nanopoulos, Ref. 92.

94. A. Masiero and R. D. Peccei, Phys. Lett. 108B, 111 (1982).

95. J. Ellis, M. K. Gaillard, D. V. Nanopoulos and S. Rudaz, Phys. Lett. 99B, 101 (1981); Nature 293, 41 (1981); Phys. Lett. 106B, 298 (1981).

96. B. Kayser and S. P. Rosen, these proceedings.

97. M. Veltman, Phys. Lett. 91B, 95 (1980); M. A. Green and M. Veltman, Nucl. Phys. B169, 137 (1979); F. Antonelli and L. Maiani, Nucl. Phys. B186, 269 (1981).

98. W. J. Marciano and A. Sirlin, Phys. Rev. D 22, 2695 (1980) and Nucl. Phys. B189, 442 (1981); S. Dawson, L. J. Hall and J. S. Hagelin, Phys. Rev. D 23, 2666 (1981); C. H. Llewellyn Smith and J. F. Wheater, Phys. Lett. 105B, 486 (1981) and Nucl. Phys. B208, 27 (1982); S. Sarantakos, A. Sirlin and W. J. Marciano, Nucl. Phys. B217, 84 (1983).

99. R. Shrock, Phys. Lett. 96B, 159 (1980); Phys. Rev. D 24, 1232, 1275 (1981).

100. D. Perret-Gallix et al., CERN PSCC proposal P 67 (1983).

101. J. H. Bardeen, B. Carter and S. Hawking, Comm. Math. Phys. 31, 181 (1973).

102. S. Hawking, Comm. Math. Phys. 43, 199 (1975); R. M. Wald, Comm. Math. Phys. 45, 9 (1975).

103. J. A. Wheeler, Relativity, Groups and Topology, ed., B. S. and C. M. deWitt (Gordon and Breach, N.Y., 1963).

104. S. Hawking, Comm. Math. Phys. 87, 395 (1982).

105. J. Ellis, J. S. Hagelin, D. V. Nanopoulos and M. Srednicki, SLAC-PUB-3134 and CERN TH-3619 (1983).

106. J. Ellis, J. S. Hagelin, D. V. Nanopoulos and K. Tamvakis, Phys. Lett. 124B, 484 (1983).

107. K. Kleinknecht, Proc. XVII Int. Conf. on High Energy Physics, London 1974, ed., J. R. Smith (S.R.C., 1974), p. III-23.

EXPERIMENTAL OUTLOOK

Bruce D. Winstein
The Enrico Fermi Institute, The University of Chicago
5640 Ellis, Chicago, IL. 60637

ABSTRACT

A summary of the prospects for exploiting an Intense Medium Energy Source of Strangeness is presented, particularly with regard to rare K decays and studies of CP nonconservation. The current status as well as new efforts at the existing laboratories are reviewed for some selected physics topics before possible future initiatives are considered.

INTRODUCTION

In principle a new Intense Medium Energy Source of Strangeness (or Kaon Factory[1,2,3]) would provide an enormous increase in sensitivity for rare processes as can be clearly seen in the following Table of Flux Comparisons:

TABLE I

Facility	K_L decays/sec
Ultimate "Factory"[4]	$\approx 10^9$
Ultimate BNL	$\approx 10^7$
Ultimate FNAL	$\approx 10^6$
Proposed BNL[5]	$\approx 10^5$

The values for the decay rates are order-of-magnitude estimates for the number of K_L decays, corrected for duty cycle, occurring in a decay volume whose length is 10% of a mean decay length; with the exception of the proposed BNL experiment where the decay region is approximately 3% of a mean decay length.

A measure of the improvement in sensitivity that can be obtained is apparent by noting that, with an apparatus with 10% acceptance, one could collect one $K_L \to 2\mu$ per sec; a total of only 30 $K_L \to 2\mu$ events have been seen with enormous efforts by several groups.

There are serious problems which will face any experiment that wants to capitalize on the higher available rate. The most serious of these is probably the load on the apparatus. With a 10% acceptance, one could sample 10^8 K_L decays per second. However this would imply a singles rate of about 10^9/sec from K_L decays alone. This already is more than two orders magnitude greater than current detectors have operated. In addition, the cleanliness of the beam is critical; what fraction of the singles rates that the detector sees are due only to K_L decays? Most previous experiments have run with this fraction being between 0.01 and 0.1. Additional problems center around the development of selective enough triggers and in the processing of vast numbers of events.

A SELECTION OF RARE K-DECAY EXPERIMENTS

Before we begin a discussion of specific decay modes, a few general observations are in order. There are in principle about five orders of magnitude between the sensitivity of previous experiments and those ultimately imaginable at a new facility. While it is usually true that, given sufficient physics interest in improving sensitivity, someone will figure out a way to improve an existing limit by an order of magnitude, it is difficult to make a step which is significantly greater. Indeed, one must grapple with the unforeseen difficulties in the handling of high rates and the rejection of backgrounds at a level ten times more sensitive than previously obtained before it is possible to contemplate an additional order of magnitude. Thus improvements of five orders of magnitude will come very slowly.[6] We also note that three of these orders are in principle obtainable at existing facilities.

$K_L \to \mu e$

The physics interest in this mode is paramount[7] and has been emphasized time and again. The current limit is somewhat ill defined: the 1971 LBL experiment[8] quotes an upper limit of 1.6×10^{-9}. They used a single spectrometer magnet and kept the material to a minimum, obtaining an extraordinary mass resolution of 1.1 MeV/c (for the $K_L \to \pi^+\pi^-$ mode). However, because the experiment did not see any $K_L \to 2\mu$ events at roughly the same sensitivity, the Particle Data Group has elected to use as a limit[9]

$$BR(K_L \to \mu e) < 6 \times 10^{-6}, \tag{1}$$

the result of the next most sensitive experiment. However, were the actual branching ratio equal to 6×10^{-6}, the LBL experiment would have had to have missed over 5000 $K_L \to \mu e$ events which is rather unlikely. Thus, we will here take the current limit to be in the range of 10^{-8}

At BNL, there is a recently approved experiment[10] to search for the existence of this mode to a level of 10^{-10}. They again use a single spectrometer magnet and will obtain a mass resolution of 1.5 MeV/c by using minidrift chambers. Resolution is of course important for distinguishing against background, the primary source being the copious $K_L \to \pi e \nu$ with the pion simulating a muon either by decay or by "punch-through".

There exists a preliminary design[11] for a first-step Factory experiment to study this decay mode. In layout, it is close to the experiment[12] of Shochet et. al. which studied the $K_L \to 2\mu$ mode: separate spectrometers track the final state particles and each has its momentum measured twice ("double bend"). Roughly, the parameters of this experiment are the following:

incident K_L's/sec	10^8
K_L decays/sec	10^7
sampled K_L decays/sec	10^5
sensitivity/10^7 sec run	10^{-12}
mass resolution	1 MeV
background	<1 event @ 10^{-12} level

As can be seen, there is some conservatism in this design. The proposed K flux is about two orders of magnitude less than that ultimately achievable, and this factor can be "traded" for cleanliness. The experiment could make use of larger targeting angles and/or an absorber such as Deuterium which preferentially removes neutrons from the beam. The acceptance (for $K_L \to \mu e$) is only 1%. Preliminary background estimates (Monte Carlo studies at the 10^{-12} level are not trivial!) indicate that 10^{-12} can be reached.

$K \to \pi\mu e$

An ambitious experiment[13] to search for $K^+ \to \pi^+ \mu^+ e^-$ at the 10^{-11} level is going to be performed at BNL. The current limit[14] for this mode is $<5 \times 10^{-9}$. Although the phase space suppression factor is significant with respect to the $K \to \mu e$ mode, here we are sensitive to a new scalar or vector interaction. The experimental signature is particularly clean as well: the incident momentum is a known 6 GeV/c and the final state consists of three charged particles emanating from a single point (within the beam). Čerenkov counters, a lead glass array, and a muon filter allow redundant particle identification to provide sufficient rejection of background the most serious of which is the decay $K^+ \to \pi^+ \pi^+ \pi^-$ where one π^+ decays and the π^- simulates an electron.

The analogous reaction $K_L \to \pi^0 e\mu$ has not been seriously investigated. The branching ratio should be greater than for the charged decay as a result of the lifetime difference. The major background would presumably again arise from the $K_L \to \pi^0 \pi^+ \pi^-$ decay with one π decaying while the other simulates an electron.

$K_L \to \pi^0 e^+ e^-$

This mode is predominantly CP-violating[15] and provides a nice test of higher-order weak effects. The current limit[16] is: $BR < 2.3 \times 10^{-6}$. This mode is particularly clean from an experimental point of view and one could very likely perform an experiment sensitive to this mode at the $10^{-10} - 10^{-11}$ level at FNAL or BNL. There exists a proto-proposal[17] for LAMPF II which claims to be able to detect 100 events of this mode at the 10^{-11} level!

$K^+ \to \pi^+$ + NONINTERACTING NEUTRALS

This mode has received considerable attention of late. It is sensitive to the number of low-mass neutrino types as well as to axion-like entities. The standard model predicts a signal at the 10^{-10} level. An experiment[18] to be performed at BNL, which could reach this level, is under discussion. It would run at $\approx 3 \times 10^5$ stopping K^+ per second in a separated beam with a nearly 4π detector to see the $\pi^+ \to \mu^+ \to e^+$ decay chain. The full acceptance would be about 1% so that a run of about a year would be required. The experiment would be sensitive to the region of π^+ momentum above the $\pi^+\pi^0$ point so that the invariant mass of the neutrals must be less than the π^0 mass. Since this would represent about a three order of magnitude improvement in the sensitivity, it is difficult at present to consider yet further improvements at a Kaon Factory. However, it is thought that for this proposed BNL experiment the limitation will be statistics so that one would profit from the availability of a better separated if not more intense beam. Finally we point out that the possibility of performing this experiment with K^+ decays in-flight is also under consideration.[19]

CP NONCONSERVATION EXPERIMENTS

In this section, we will discuss some experiments to probe CP nonconservation which are currently being carried out or proposed at existing labs. We will discuss where appropriate some considerations relevant to extending these studies at a Kaon Factory.

TRANSVERSE MUON POLARIZATION IN $K\mu_3$ DECAY

This experiment has been performed at BNL in both K_L and K^+ decay by the same group.[20] Combining the results from both experiments, they find $P_n = (-1.9 \pm 3.6) \times 10^{-3}$ where P_n is the measured polarization of the muon perpendicular to the plane defined by $P\pi$ and $P\mu$. In terms of ξ, the ratio of form factors in $K\mu_3$ decay, this result implies that $\mathrm{Im}\xi = -0.01 \pm 0.02$. The experiment collected about 20×10^6 events but was still statistics limited. It would be desirable to push the sensitivity still further as a positive effect would signal a mechanism other than the "standard" KM model. This possibility has been considered[21] in the context of a Kaon Factory; it is concluded that a significant improvement is attainable by capitalizing upon a separated and more intense K^+ beam.

η_{+-o}

This parameter of CP nonconservation is expected to be equal to η_{+-} in the "superweak" model, and very nearly equal to η_{+-} in the "standard" KM model. At present, we know very little about this parameter:[22] $|\eta_{+-o}|^2 < 0.12$.

An experiment[23] is planned to run at Fermilab which should considerably improve our knowledge of η_{+-o}. The expected error on η_{+-o} is given as $\eta_{+-}/4$ or $\simeq 6 \times 10^{-4}$. The branching ratio for $K_S \to \pi^+\pi^-\pi^0$ is expected to be $\approx 10^{-9}$ in the standard (or superweak) model so that a direct measurement is not feasible. The experiment exploits the interference between the K_L and K_S decay in an initially pure K^0 beam. Schematically, one expects constructive interference at early times and destructive interference at later times as shown in the figure:

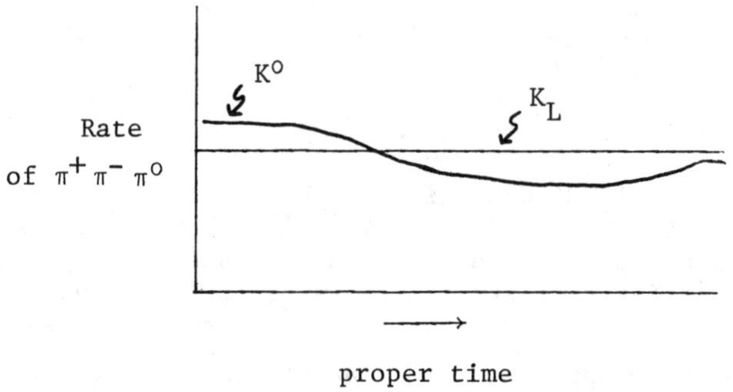

The expected asymmetry between upstream and downstream decays is only about 1×10^{-3} if $\eta_{+-o} = \eta_{+-}$. The detection efficiency is calibrated with a K_L beam (i.e. the target is far upstream). The expected asymmetry, with its error, can then be crudely represented as

$$a = \frac{N_o^U}{N_L^U} - \frac{N_o^D}{N_L^D} = (1 \pm 0.25) \times 10^{-3} \tag{2}$$

where $N_{o(L)}^{U(D)}$ represents the detected number of events either up or downstream from the K^0 or K_L beams. Thus, as a measure of the difficulty of the experiment, we see that we need about 60×10^6 in each sample with systematics controlled to that level.

DIRECT (MILLIWEAK) CP NONCONSERVATION

We consider some experiments designed to directly observe possible differences between particle and antiparticle decay amplitudes. Such effects are governed by the parameter ε' which in the KM model, is predicted[24] to be bigger than about $.002 \times \varepsilon$ where ε is the mass mixing parameter ($|\varepsilon| \approx 2 \times 10^{-3}$). Presently, one knows[25] that $|\varepsilon'/\varepsilon|$ is less than about 0.02.

At early times (before mixing takes place) one could detect a milliweak effect in the kaon system[26] by directly measuring the decay rate difference between K^0 and $\bar{K}^0$ to $\pi^+\pi^-$. Specifically,

$$\frac{\Gamma(K^0 \to 2\pi)}{\Gamma(\bar{K}^0 \to 2\pi)} - 1 = 4\varepsilon' \ . \qquad (3)$$

The advantage of this method, that one need not measure the $2\pi^0$ mode to explore the milliweak interaction, does not outweigh the problem of the extremely small effect. A similar effect has been calculated[27] for Λ decay:

$$\frac{\Gamma(\Lambda \to p\pi^-)}{\Gamma(\bar{\Lambda} \to \bar{p}\pi^+)} - 1 \sim 0(10^{-1})\,\varepsilon' \qquad (4)$$

which again looks too small to be directly seen.

ε' MEASUREMENT AT LEAR

An experiment to measure ε' using kaons produced from $\bar{p}p$ annihilations at rest at the LEAR facility in Geneva has been discussed.[28] An attractive feature is the high rate of tagged kaons: it is estimated that through the reaction $\bar{p}p \to K^\pm \pi^\mp \{\frac{K^0}{\bar{K}^0}\}$ about 2×10^8 tagged K^0's are produced per day. One can then directly compare the time evolution of initially pure K^0's or $\bar{K}^0$'s.

The authors propose that the integrated asymmetry in $\pi^0\pi^0$ decays be measured. Specifically, if $N_o(\bar{N}_o)$ is the number of observed $K^0 \to 2\pi^0$ ($\bar{K}^0 \to 2\pi^0$) events integrated over about 20 K_S lifetimes, then the following can be shown:

$$A^{oo} \equiv (N_o - \bar{N}_o)/(N_o + \bar{N}_o) \simeq 4\,\mathrm{Re}\,\eta_{oo} - 2\,\mathrm{Re}\,\varepsilon \ . \qquad (5)$$

Now, since $\eta_{oo} = \varepsilon - \varepsilon'$, we have

$$A^{oo} = 2\,\mathrm{Re}\,\varepsilon - 8\,\mathrm{Re}\,\varepsilon' \qquad (6)$$

so that the integrated particle/antiparticle asymmetry, while dominated by ε, still is somewhat sensitive to ε'. Since $\eta_{+-} = \varepsilon + \varepsilon'$, we obtain

$$(\varepsilon'/\varepsilon) = 0.1\ (2 - A^{oo}/\mathrm{Re}\ \eta_{+-}) \tag{7}$$

Then, to have an error in ε'/ε of .001, which we will see is the goal of the next round of "conventional" experiments, one would require that N_o, $\bar{N}_o$ be on the order of 10^9. While this appears to be attainable in a few weeks of running, the real challenge would be the control of systematics to the required level.

It appears to the author that the LEAR facility might be the ideal place to observe the parameter η_{+-o}. Specifically, the sign of the interference term between $K_L \to \pi^+ \pi^- \pi^0$ and $K_S \to \pi^+ \pi^- \pi^0$ is reversed for K^0 and $\bar{K}^0$. Thus the difference between the two time distributions ought to amplify the expected asymmetry while the sum of the two effectively measures the detector efficiency vs. proper time. The signature is clean even without an elaborate electromagnetic detector. The expected high acceptance at early proper times - where the asymmetry is greatest - contrasts with the method of using a K^0 beam so that one should readily be able to establish that $\eta_{+-o} \approx \eta_{+-}$.

ε' MEASUREMENTS AT FNAL, BNL, AND THE CERN SPS

In June of 1982, Fermilab experiment (E617)[29] completed its data taking and analysis is still proceeding. The experiment determines ε'/ε by a measurement of $\left|\frac{\eta_{+-}}{\eta_{oo}}\right|^2$. Each of the two modes (charged and neutral) of $K \to 2\pi$ is studied independently. For each study, a dual beam arrangement is employed: in one of the beams is placed a regenerator so the $K_L \to 2\pi$ and $K_S \to 2\pi$ decay rates are measured simultaneously. The identical regenerator (which alternates between the beams) is used for both charged and neutral modes.

The experiment collected about 3500 $K_L \to 2\pi^0$ decays in a large lead glass array. The expected sensitivity to ε'/ε is about ± 0.004; systematics need to be controlled to about the 1.5% level so that statistical errors dominate.

An experiment[30] to study ε' is currently running at BNL. Its goals are similar to those of the FNAL experiment with, however, a different method. Both $\pi^+ \pi^-$ and $\pi^0 \pi^0$ modes are detected simultaneously (in different solid angles) first in vacuum, then with a regenerator. The experimenters are also attempting to make use of the possibly very different interference patterns downstream of the thick regenerator between $\pi^+ \pi^-$ and $\pi^0 \pi^0$ as a signal of a non-zero ε'.

A major CERN experiment[31] devoted to studying ε'/ε is currently in the construction phase. While both the above efforts rely upon a magnetic spectrometer for charged particle momentum determination, the CERN effort has no magnetic field. Instead, they use wire chambers and a large liquid Argon electromagnetic calorimeter followed by a hadronic calorimeter. As well, they do not use a regenerator to provide K_S decays. Instead they are simultaneously sensitive to $\pi^+\pi^-$ and $2\pi^0$ modes and run with a far target for K_L decays and a near one for K_S decays. The experiment should run in 1985 and collect at least 10^5 $K_L \to 2\pi^0$ events.

Finally, a new version of the FNAL experiment has been proposed[32] for the Tevatron. It will use the same double beam technique but will require a larger aperture magnet and chambers. It makes use of a new cleaner K_L beam and takes advantage both of the higher energy and the long spill at the Tevatron. The goal of this effort is at least 10^5 $K_L \to 2\pi^0$ decays and a precision on ε'/ε of .001; systematics for this as well as the CERN experiment must be held to $< 1/2\%$.

ε'/ε AT A KAON FACTORY

To evaluate the experimental need for ε'/ε studies at a Kaon Factory, one must surely await the results of the current and next generation accelerator experiments. Indeed the expected sensitivities of these efforts are at or below the level that is predicted by at least the standard model. Nevertheless, it is obvious that one may need and desire to push even further than .001 precision (on ε'/ε) so we may speculate on how that could be accomplished.

It is doubtful that further improvements could come from the existing accelerators: one is already straining a bit both with respect to rate and systematics.[33] It would be desirable to have an "amplifier".

Such an amplifier has been suggested by Adair.[34] Downstream of a thick regenerator, the unscattered 2π decay amplitude has a contribution from $K_L \to 2\pi$ as well as from $K_S \to 2\pi$. Because of the $K_{L,S}$ mass difference, the well known time-dependent interference pattern results. The point is that this pattern depends solely upon ρ/η where ρ is the regeneration amplitude and η is either $\eta_{+-} = \eta_{oo}$ depending upon which mode is studied. Thus, in the superweak model where $\eta_{+-} = \eta_{oo}$, the patterns should be identical and any difference, then, is a signal of a nonzero ε'. By careful choice of the parameters of the regenerator, one can have a very large difference exactly in the region of the interference minimum where one has nearly a pure K_2 CP eigenstate. For coherent regeneration, the minimum occurs at about 9 lifetimes so that as a result, the event rate is very small. However, if one could take advantage of the available rate at a Kaon Factory, then a study by the author indicates that it could be possible to improve the sensitivity to ε'/ε about a factor of 3, to about 3×10^{-4}, without requiring even better control of systematics. In addition, one might consider using diffraction regeneration where the interference minimum occurs at about 3 lifetimes.

There are other features of a K_L beam at a Kaon Factory that could prove desirable. First, one could in fact make a monochromatic[35] K_L beam with about the same intensity as is now available (broad-band) elsewhere. This would be desirable in that momentum dependent systematics would be reduced as would background (an extra constraint would be available). Second, one could make smaller and cleaner beams. Finally one might consider using the π^0 Dalitz decay in the study of the $2\pi^0$ mode: $K_{L(S)} \to \pi^0 \pi^0 \to 2\gamma + \gamma e^+ e^-$: this would provide a far better determination of the K decay vertex and as a result would provide another tool for background suppression.

CONCLUSION

The actual scenario for the exploitation of the 100-fold increase in available kaon flux either in CP nonconservation or rare decay studies must clearly await the experience from new initiatives at current accelerators. Nevertheless it is clear that such a facility would provide a degree of flexibility to the experimenters not currently available at existing machines, and, possibly a glimpse at the next level of substructure.

ACKNOWLEDGEMENTS

The author is pleased to acknowledge useful interactions with M. Campbell, D. Carlsmith, W. Louis, L. Littenberg and K. Nishikawa.

REFERENCES

1. C.M. Hoffman, these proceedings.
2. D. Bryman, these proceedings.
3. L.S. Littenberg, these proceedings.
4. "Possible LAMPF II Fluxes", H.A. Thiessen, Los Alamos memorandum, MP-10.
5. See reference 13 where the proposed K^+ decay rate is $\approx 10^6$.
6. The author would not object to being shown wrong on this point.
7. See, for example, O. Shanker, Nucl. Phys. B206, 253 (1982); S. Dimopoulos, S. Raby, and G.L. Kane, Nucl. Phys. B182, 77 (1981) and references therein.
8. A.R. Clark et. al., Phys. Rev. Lett. 26, 1667 (1971).
9. V. Fitch, R.F. Roth, J. Russ, and W. Vernon, Phys. Rev. 164, 1711 (1967).
10. AGS experiment No. 780, BNL-Yale collaboration. (M.P. Schmidt/ W.M. Morse spokesmen.)
11. H. Lubatti, private communication.
12. M.J. Shochet et. al., Phys. Rev. D19, 1965 (1979).
13. AGS experiment No. 777, BNL-Pittsburgh-Washington-Yale collaboration. (M.E. Zeller spokesman.)
14. A.M. Diamant-Berger, et. al., Phys. Lett. 62B, 485 (1976).

15. M.K. Gaillard and B.W. Lee, Phys. Rev. D10, 897 (1974).
16. A.S. Carroll et. al., Phys. Rev. Lett. 44, 525 (1980).
17. "An Experiment to Measure the Decay $K_L \rightarrow \pi^0 e^+ e^-$ at LAMPF II," L. Littenberg, BNL.
18. W. Louis and L. Littenberg, Private communication.
19. J.W. Cronin, private communication; S. Wojcicki, private communication.
20. M.K. Campbell et. al., Phys. Rev. Lett. 47, 1032 (1981).
21. "The CP-Violating Polarization in $K^+\mu_3$ and $K^0\mu_3$ Decays", Ted Kalogeropoulos, Temple University.
22. M. Metcalf et. al., Phys. Lett. 40B, 703 (1972).
23. FNAL experiment E621, Rutgers-Michigan-Minnesota-Wisconsin collaboration. (G. Thomson spokesman.)
24. F. Gilman and J.S. Hagelin, SLAC PUB-3087.
25. M. Holder et. al., Phys. Lett. 40B, 141 (1972); M. Banner et. al., Phys. Rev. Lett. 28, 1597 (1972).
26. "Experiments in the Kaon System", B. Winstein, Proceedings of the Los Alamos Workshop "Nuclear and Particle Physics at Energies Below 31 GeV: New and Future Aspects", 1981, LA 8775-C, pp. 18-35.
27. L. Wolfenstein, private communication.
28. "Strong and Weak CP Violation at LEAR", E. Gabathuler and P. Pavlopoulos, CERN-EP/82-118 (1982).
29. FNAL experiment E617, Chicago-Saclay collaboration. (B. Winstein spokesman.)
30. AGS experiment #749 BNL-Yale collaboration. (M.P. Schmidt/W.M. Morse spokesmen.)
31. CERN-Dortmund-Orsay-Pisa-Siegen collaboration. (H. Wahl spokesman.)
32. FNAL experiment P731, Chicago-Saclay collaboration. (B. Winstein spokesman.)
33. Again the author would like to be proven wrong on this point.
34. See reference 30 and reference 26.
35. "A Monochromatic Neutral Kaon Beam", C. Hoffman, LA-8444-MS (1980).

WHAT IS ALL THE TSIMESS ABOUT?

S. P. Rosen
Purdue University, West Lafayette, Indiana 47909

In his kind introduction, Joel Primack observed that I was about to pass from the land of disenchantment to the land of enchantment, from Washington, D.C. to New Mexico; but he neglected to mention that, on the way, I have to pass through Santa Cruz, a land which few words can describe.

Having arrived here early on Friday afternoon, I decided to walk around and explore the town before the evening TSIMESS began. Just out of the motel, I almost collided with a man in brown pyjamas, thongs, and a baseball hat. He wore a beard fit for the Supreme Orthodox Metropolitan of Moscow and carried a bedroll under his arm. As I walked behind him down the street, he would glance back from time to time as if to say: "Don't follow too closely!" Another character of the same ilk stepped out of a side street, but to my surprise, no words were exchanged. The Metropolitans must have taken me for a government man in my conventional jacket and tie, and they preferred to give no hint of mutual recognition.

Eventually my Metropolitan led me to the bus station, a place where time has stood still for fifteen years. It belongs to the era of hippies and flower children, the stoned and the supine. One person panhandled me, and another urged me to tune in to the Saturn Cafe, where the mud pies send you out of this world, and the bathrooms set you afloat in black light. I looked around for the Metropolitan to guide me, but he had vanished in the evening crowd. It was time to return to the real world -- rare decays and technicolor, massive neutrinos and grand unification, gravity and gravitinos.

I asked to speak at this banquet in order to comment on some of the questions raised at the plenary session today. This need not be an era of zero-sum games and cutthroat competition between different subfields of physics! After all, it is by no means clear that major advances in particle physics will be made exclusively through the traditional channel of higher and higher energies and without the active assistance of nuclear physics. No, indeed far from it!

Let me state clearly that the views I am about to express are entirely my own, and not necessarily those of any organization or institution with which I am associated. When I came to Washington in October 1981, we had to prepare for a threatened cut of roughly 16% in the budget. Today we look forward, Congress willing, to an 18% increase. Why, you may well ask, this complete reversal of direction in science funding?

As a bystander on the Washington scene, I would venture the opinion that, in the last year, both the Administration and Congress have come to realize that, during the past decade, we have allowed the health of the scientific enterprise in the United States to deteriorate to a point which threatens the economic vitality and general security of the nation. We are not facing imminent catastrophe, but the danger is there in the shadows. The Administration and

0094-243X/83/1020235-02 $3.00

the Congress, Republicans and Democrats, all seem to agree that "we must do something about it."

Given this "sea change" in the general climate, I think that the physics community has a rare opportunity to gain support for major projects for the decade of the nineties. The opportunity exists for every subfield that can put forward an exceptionally strong scientific case for a project or plan which has broad support within its own community. It will require discipline, in that subfields will have to establish firm scientific priorities; and it will require statesmanship, in that individuals will have to look to the good of the field as a whole, rather than to their own narrow interests. In this context, I think that the High Energy Physics Community has an opportunity to gain support for a long range plan, but it must present the strongest possible scientific case and a united front. The same holds separately and independently for medium energy and nuclear physics.

Let me caution you that opportunities do not last for ever, and that the process of government tends to soften the extremes. The cut in October 1981 ended up at 12% instead of 16%, and today's increase of 18% will likely be somewhat lower when it actually comes into effect. So much then, for the observations of a bystander.

In conclusion, I would like to thank our hosts at the Santa Cruz Institute for Particle Physics for their excellent enthusiasm, organization, and hospitality. Deserving special mention are the ladies, Georgia and Candace who have bestowed upon us the labor of their hands and the warmth of their smiles. Thank you very much.

CONFERENCE PROGRAM

Saturday, March 19, 1983

Welcome/Statement of Purpose
T. Goldman

LAMPF II
C. M. Hoffman

TRIUMF Kaon Factory
D. Bryman

Brookhaven Kaon Factory
L. Littenberg

Hyperon and Hypernuclear Physics with Intense Beams
B. F. Gibson

Theoretical Issues Within the Standard Model
M. K. Gaillard

Theoretical Issues Outside the Standard Model
F. Wilczek

Organization and First Meeting of Working Groups

Supper - Kresge Town Hall, Kresge College

Second Meeting of Working Groups

Sunday, March 20, 1983

Final Meetings of Working Groups

Report of Discrete Symmetries (CPT) Working Group
L. Wolfenstein, Chairman

Report of Precision Weak Interactions Working Group
R. Shrock, Chairman

Report of New Particles and Properties Working Group
G. Kane, Chairman

Summary: Theoretical Perspectives
J. Ellis

Experimental Outlook
B. Winstein

Banquet

Monday, March 21, 1983

Working Group Chairpersons and Scientific Secretaries Draft Reports.

CONFERENCE PARTICIPANTS

Georges Azuelos
University of Montreal

Stephen Barr
University of Washington

Greg Beall
University of Oregon

Ikaros Bigi
Institut fur Theoretische Physik
Aachen, West Germany

Gustavo Branco
Virginia Polytechnic Institute
and State University

Richard Brower
University of California, Santa Cruz

Doug Bryman
University of British Columbia

Robert N. Cahn
Lawrence Berkeley Laboratory

David O. Caldwell
University of California, Santa Barbara

Jean-Luc Cambier
University of California, Santa Cruz

Roger Carlini
Los Alamos National Laboratory

Darwin Chang
Carnegie-Mellon University

T. P. Cheng
Lawrence Berkeley Laboratory

Farhad Daghighian
University of California, Irvine

N. Deshpande
University of Oregon

David Dorfan
University of California, Santa Cruz

Keith Dow
University of California, Santa Cruz

Monica Eder
Zurich, Switzerland

Pat Egan
Yale University

John Ellis
Stanford Linear Accelerator Center

Ralph Fabrizio
University of California, Santa Cruz

M. Ferro-Luzzi
CERN/Geneva, Switzerland

Ricardo Flores
University of California, Santa Cruz

Melissa Franklin
Lawrence Berkeley Laboratory

Jean-Marie Frere
University of Michigan

Mary K. Gaillard
Lawrence Berkeley Laboratory

M. B. Gavela
Brandeis University

Benjamin F. Gibson
Los Alamos National Laboratory

Frederick J. Gilman
Stanford Linear Accelerator Center

Terry Goldman
Los Alamos National Laboratory

Kim Griest
University of California, Santa Cruz

Howard E. Haber
University of California, Santa Cruz

John S. Hagelin
Stanford Linear Accelerator Center

Osamu Hashimoto
INS/Tokyo

Wick Haxton
Los Alamos National Laboratory

Doug Hellinger
University of California, Santa Cruz

Peter Herczeg
Los Alamos National Laboratory

Clemens Heusch
University of California, Santa Cruz

C. M. Hoffman
Los Alamos National Laboratory

D. Jensen
University of Massachusetts

Pat Kalyniak
University of British Columbia

A. N. Kamal
University of Alberta

Gordon L. Kane
University of Michigan

Boris Kayser
National Science Foundation

Joe Lach
Fermilab

John Lazenby
University of California, Santa Cruz

Ling-Fong Li
Carnegie-Mellon University

V. Gordon Lind
Utah State University

L. Littenberg
Brookhaven National Laboratory

William Louis III
Princeton University

Henry Lubatti
University of Washington

Maurizio Lusignoli
Istito di Fisica/Roma, Italy

R. Macek
Los Alamos National Laboratory

Fesseha G. Mariam
Los Alamos National Laboratory

R. E. Mischke
Los Alamos National Laboratory

Rabindra Mohapatra
University of Maryland

Darragh Nagle
Los Alamos National Laboratory

Michael Nauenberg
University of California, Santa Barbara

John Ng
University of British Columbia

Takamitsu Oka
Los Alamos National Laboratory

A. Olin
University of British Columbia

Michael Peskin
Stanford Linear Accelerator Center

Charles Picciotto
University of Victoria

Bernard Pire
Stanford Linear Accelerator Center

H. David Politzer
California Institute of Technology

Joel Primack
University of California, Santa Cruz

Halsie Reno
Stanford Linear Accelerator Center

S. Peter Rosen
National Science Foundation

Hartmut F.-W. Sadrozinski
University of California, Santa Cruz

V. Sandberg
Los Alamos National Laboratory

Gary Sanders
Los Alamos National Laboratory

Mike Schmidt
Yale University

Howard Schnitzer
Brandeis University

Abraham Seiden
University of California, Santa Cruz

Goran Senjanović
Brookhaven National Laboratory

Stephen Sharpe
Lawrence Berkeley Laboratory

Marc Sher
University of California, Irvine

Robert Shrock
SUNY-Stony Brook

Jerry Stephenson
Los Alamos National Laboratory

Mark Strovink
Lawrence Berkeley Laboratory

Richard Talaga
University of Maryland

Giora Tarnopolsky
Stanford Linear Accelerator Center

Peter Vogel
California Institute of Technology

Frank Wilczek
University of California, Santa Barbara

Bruce Winstein
University of Chicago

Mike Witherell
University of California, Santa Barbara

Lincoln Wolfenstein
Carnegie-Mellon University

T. Yamazaki
KEK
Ibaraki-ken, Japan

Arnulfo Zepeda
CIEA-IPN
Mexico

AIP Conference Proceedings

		L.C. Number	ISBN
No.1	Feedback and Dynamic Control of Plasmas	70-141596	0-88318-100-2
No.2	Particles and Fields - 1971 (Rochester)	71-184662	0-88318-101-0
No.3	Thermal Expansion - 1971 (Corning)	72-76970	0-88318-102-9
No.4	Superconductivity in d-and f-Band Metals (Rochester, 1971)	74-18879	0-88318-103-7
No.5	Magnetism and Magnetic Materials - 1971 (2 parts) (Chicago)	59-2468	0-88318-104-5
No.6	Particle Physics (Irvine, 1971)	72-81239	0-88318-105-3
No.7	Exploring the History of Nuclear Physics	72-81883	0-88318-106-1
No.8	Experimental Meson Spectroscopy - 1972	72-88226	0-88318-107-X
No.9	Cyclotrons - 1972 (Vancouver)	72-92798	0-88318-108-8
No.10	Magnetism and Magnetic Materials - 1972	72-623469	0-88318-109-6
No.11	Transport Phenomena - 1973 (Brown University Conference)	73-80682	0-88318-110-X
No.12	Experiments on High Energy Particle Collisions - 1973 (Vanderbilt Conference)	73-81705	0-88318-111-8
No.13	π-π Scattering - 1973 (Tallahassee Conference)	73-81704	0-88318-112-6
No.14	Particles and Fields - 1973 (APS/DPF Berkeley)	73-91923	0-88318-113-4
No.15	High Energy Collisions - 1973 (Stony Brook)	73-92324	0-88318-114-2
No.16	Causality and Physical Theories (Wayne State University, 1973)	73-93420	0-88318-115-0
No.17	Thermal Expansion - 1973 (lake of the Ozarks)	73-94415	0-88318-116-9
No.18	Magnetism and Magnetic Materials - 1973 (2 parts) (Boston)	59-2468	0-88318-117-7
No.19	Physics and the Energy Problem - 1974 (APS Chicago)	73-94416	0-88318-118-5
No.20	Tetrahedrally Bonded Amorphous Semiconductors (Yorktown Heights, 1974)	74-80145	0-88318-119-3
No.21	Experimental Meson Spectroscopy - 1974 (Boston)	74-82628	0-88318-120-7
No.22	Neutrinos - 1974 (Philadelphia)	74-82413	0-88318-121-5
No.23	Particles and Fields - 1974 (APS/DPF Williamsburg)	74-27575	0-88318-122-3
No.24	Magnetism and Magnetic Materials - 1974 (20th Annual Conference, San Francisco)	75-2647	0-88318-123-1
No.25	Efficient Use of Energy (The APS Studies on the Technical Aspects of the More Efficient Use of Energy)	75-18227	0-88318-124-X

No.26	High-Energy Physics and Nuclear Structure - 1975 (Santa Fe and Los Alamos)	75-26411	0-88318-125-8
No.27	Topics in Statistical Mechanics and Biophysics: A Memorial to Julius L. Jackson (Wayne State University, 1975)	75-36309	0-88318-126-6
No.28	Physics and Our World: A Symposium in Honor of Victor F. Weisskopf (M.I.T., 1974)	76-7207	0-88318-127-4
No.29	Magnetism and Magnetic Materials - 1975 (21st Annual Conference, Philadelphia)	76-10931	0-88318-128-2
No.30	Particle Searches and Discoveries - 1976 (Vanderbilt Conference)	76-19949	0-88318-129-0
No.31	Structure and Excitations of Amorphous Solids (Williamsburg, VA., 1976)	76-22279	0-88318-130-4
No.32	Materials Technology - 1976 (APS New York Meeting)	76-27967	0-88318-131-2
No.33	Meson-Nuclear Physics - 1976 (Carnegie-Mellon Conference)	76-26811	0-88318-132-0
No.34	Magnetism and Magnetic Materials - 1976 (Joint MMM-Intermag Conference, Pittsburgh)	76-47106	0-88318-133-9
No.35	High Energy Physics with Polarized Beams and Targets (Argonne, 1976)	76-50181	0-88318-134-7
No.36	Momentum Wave Functions - 1976 (Indiana University)	77-82145	0-88318-135-5
No.37	Weak Interaction Physics - 1977 (Indiana University)	77-83344	0-88318-136-3
No.38	Workshop on New Directions in Mossbauer Spectroscopy (Argonne, 1977)	77-90635	0-88318-137-1
No.39	Physics Careers, Employment and Education (Penn State, 1977)	77-94053	0-88318-138-X
No.40	Electrical Transport and Optical Properties of Inhomogeneous Media (Ohio State University, 1977)	78-54319	0-88318-139-8
No.41	Nucleon-Nucleon Interactions - 1977 (Vancouver)	78-54249	0-88318-140-1
No.42	Higher Energy Polarized Proton Beams (Ann Arbor, 1977)	78-55682	0-88318-141-X
No.43	Particles and Fields - 1977 (APS/DPF, Argonne)	78-55683	0-88318-142-8
No.44	Future Trends in Superconductive Electronics (Charlottesville, 1978)	77-9240	0-88318-143-6
No.45	New Results in High Energy Physics - 1978 (Vanderbilt Conference)	78-67196	0-88318-144-4
No.46	Topics in Nonlinear Dynamics (La Jolla Institute)	78-057870	0-88318-145-2
No.47	Clustering Aspects of Nuclear Structure and Nuclear Reactions (Winnepeg, 1978)	78-64942	0-88318-146-0
No.48	Current Trends in the Theory of Fields (Tallahassee, 1978)	78-72948	0-88318-147-9
No.49	Cosmic Rays and Particle Physics - 1978 (Bartol Conference)	79-50489	0-88318-148-7

No. 50	Laser-Solid Interactions and Laser Processing - 1978 (Boston)	79-51564	0-88318-149-5
No. 51	High Energy Physics with Polarized Beams and Polarized Targets (Argonne, 1978)	79-64565	0-88318-150-9
No. 52	Long-Distance Neutrino Detection - 1978 (C.L. Cowan Memorial Symposium)	79-52078	0-88318-151-7
No. 53	Modulated Structures - 1979 (Kailua Kona, Hawaii)	79-53846	0-88318-152-5
No. 54	Meson-Nuclear Physics - 1979 (Houston)	79-53978	0-88318-153-3
No. 55	Quantum Chromodynamics (La Jolla, 1978)	79-54969	0-88318-154-1
No. 56	Particle Acceleration Mechanisms in Astrophysics (La Jolla, 1979)	79-55844	0-88318-155-X
No. 57	Nonlinear Dynamics and the Beam-Beam Interaction (Brookhaven, 1979)	79-57341	0-88318-156-8
No. 58	Inhomogeneous Superconductors - 1979 (Berkeley Springs, W.V.)	79-57620	0-88318-157-6
No. 59	Particles and Fields - 1979 (APS/DPF Montreal)	80-66631	0-88318-158-4
No. 60	History of the ZGS (Argonne, 1979)	80-67694	0-88318-159-2
No. 61	Aspects of the Kinetics and Dynamics of Surface Reactions (La Jolla Institute, 1979)	80-68004	0-88318-160-6
No. 62	High Energy e^+e^- Interactions (Vanderbilt , 1980)	80-53377	0-88318-161-4
No. 63	Supernovae Spectra (La Jolla, 1980)	80-70019	0-88318-162-2
No. 64	Laboratory EXAFS Facilities - 1980 (Univ. of Washington)	80-70579	0-88318-163-0
No. 65	Optics in Four Dimensions - 1980 (ICO, Ensenada)	80-70771	0-88318-164-9
No. 66	Physics in the Automotive Industry - 1980 (APS/AAPT Topical Conference)	80-70987	0-88318-165-7
No. 67	Experimental Meson Spectroscopy - 1980 (Sixth International Conference , Brookhaven)	80-71123	0-88318-166-5
No. 68	High Energy Physics - 1980 (XX International Conference, Madison)	81-65032	0-88318-167-3
No. 69	Polarization Phenomena in Nuclear Physics - 1980 (Fifth International Symposium, Santa Fe)	81-65107	0-88318-168-1
No. 70	Chemistry and Physics of Coal Utilization - 1980 (APS, Morgantown)	81-65106	0-88318-169-X
No. 71	Group Theory and its Applications in Physics - 1980 (Latin American School of Physics, Mexico City)	81-66132	0-88318-170-3
No. 72	Weak Interactions as a Probe of Unification (Virginia Polytechnic Institute - 1980)	81-67184	0-88318-171-1
No. 73	Tetrahedrally Bonded Amorphous Semiconductors (Carefree, Arizona, 1981)	81-67419	0-88318-172-X
No. 74	Perturbative Quantum Chromodynamics (Tallahassee, 1981)	81-70372	0-88318-173-8

No. 75	Low Energy X-ray Diagnostics-1981 (Monterey)	81-69841	0-88318-174-6
No. 76	Nonlinear Properties of Internal Waves (La Jolla Institute, 1981)	81-71062	0-88318-175-4
No. 77	Gamma Ray Transients and Related Astrophysical Phenomena (La Jolla Institute, 1981)	81-71543	0-88318-176-2
No. 78	Shock Waves in Condensed Matter - 1981 (Menlo Park)	82-70014	0-88318-177-0
No. 79	Pion Production and Absorption in Nuclei - 1981 (Indiana University Cyclotron Facility)	82-70678	0-88318-178-9
No. 80	Polarized Proton Ion Sources (Ann Arbor, 1981)	82-71025	0-88318-179-7
No. 81	Particles and Fields - 1981: Testing the Standard Model (APS/DPF, Santa Cruz)	82-71156	0-88318-180-0
No. 82	Interpretation of Climate and Photochemical Models, Ozone and Temperature Measurements (La Jolla Institute, 1981)	82-071345	0-88318-181-9
No. 83	The Galactic Center (Cal. Inst. of Tech., 1982)	82-071635	0-88318-182-7
No. 84	Physics in the Steel Industry (APS.AISI, Lehigh University, 1981)	82-072033	0-88318-183-5
No. 85	Proton-Antiproton Collider Physics - 1981 (Madison, Wisconsin)	82-072141	0-88318-184-3
No. 86	Momentum Wave Functions - 1982 (Adelaide, Australia)	82-072375	0-88318-185-1
No. 87	Physics of High Energy Particle Accelerators (Fermilab Summer School, 1981)	82-072421	0-88318-186-X
No. 88	Mathematical Methods in Hydrodynamics and Integrability in Dynamical Systems (La Jolla Institute, 1981)	82-072462	0-88318-187-8
No. 89	Neutron Scattering - 1981 (Argonne National Laboratory)	82-073094	0-88318-188-6
No. 90	Laser Techniques for Extreme Ultraviolt Spectroscopy (Boulder, 1982)	82-073205	0-88318-189-4
No. 91	Laser Acceleration of Particles (Los Alamos, 1982)	82-073361	0-88318-190-8
No. 92	The State of Particle Accelerators and High Energy Physics(Fermilab, 1981)	82-073861	0-88318-191-6
No. 93	Novel Results in Particle Physics (Vanderbilt, 1982)	82-73954	0-88318-192-4
No. 94	X-Ray and Atomic Inner-Shell Physics-1982 (International Conference, U. of Oregon)	82-74075	0-88318-193-2
No. 95	High Energy Spin Physics - 1982 (Brookhaven National Laboratory)	83-70154	0-88318-194-0
No. 96	Science Underground (Los Alamos, 1982)	83-70377	0-88318-195-9

97	The Interaction Between Medium Energy Nucleons in Nuclei-1982 (Indiana University)	83-70649	0-88318-196-7
98	Particles and Fields - 1982 (APS/DPF University of Maryland)	83-70807	0-88318-197-5
99	Neutrino Mass and Gauge Structure of Weak Interactions (Telemark, 1982)	83-71072	0-88318-198-3
100	Excimer Lasers - 1983 (OSA, Lake Tahoe, Nevada)	83-71437	0-88318-199-1
101	Positron-Electron Pairs in Astrophysics (Goddard Space Flight Center, 1983)	83-71926	0-88318-200-9
102	Intense Medium Energy Sources of Strangeness (UC-Santa Cruz, 1983)	83-72261	0-88318-201-7